EXECUTIVE SUMMARY

Marine Strategy for Torres Strait (MaSTS)

Purpose of MaSTS

The purpose of the Marine Strategy for Torres Strait (MaSTS) was defined at a first consensus-building workshop in September, 1992:

> to establish a comprehensive management framework which will permit optimum exploitation of the region's limited resource base, consistent with the needs of indigenous Torres Strait Islanders, sustainable development and minimal environmental disturbance.

The need for MaSTS

Islanders seek a positive approach to marine issues. It has seemed to us that most recent approaches proposed in Australia have been negative responses to problems which have got out of hand. Furthermore, reports and inquiries have seemed to view coastal and marine issues as occasional and exceptional side effects of development. We, on the other hand, view marine and coastal management as central not only to our livelihoods and daily food resources, but to our social and cultural environment and survival as a unique regional community.

A thorough change in approach is needed. MaSTS is an attempt by the Island Coordinating Council of Torres Strait (ICC) to carry out such a change.

> A comprehensive conservation and sustainable development strategy is needed as the foundation for future public policy and economic development in the Torres Strait region. A Marine Strategy for Torres Strait (MaSTS) would be the largest element in this approach, building on the unique relationship between Islanders and the sea. This relationship is as important for cultural traditions and social forms as for our livelihoods and future well-being (Principles and Objectives for the Future of Torres Strait, May 16, 1991, 10, corrected version).

Principles of MaSTS

MaSTS is guided by the following principles, both in its development and implementation phases:

1) The natural resources of the Torres Strait are not unlimited: there are limits to the marine environment as a pollution sink and as a provider of renewable resources. There is, therefore, a need to protect the marine environment and its resources for present and future generations.

2) Effective resource and ecosystem management requires that local residents play a leading role in decision-making processes. Many current policies and programs involving the management, protection and utilisation of the Torres Strait marine environment will need to be amended to provide for the greater involvement and representation of Torres Strait Islanders.

3) Many current policies and administrative practices are characterised by complex and overlapping jurisdictional responsibilities. There is a need to coordinate and rationalise existing arrangements and develop a comprehensive regional approach to the management of the Torres Strait marine environment.

4) The Torres Strait economy is based in large part on natural renewable marine resources. The needs of the Torres Strait Islanders, as the principal users and dependants on the marine environment, must receive priority in any competition for these resources.

5) Torres Strait Islanders have legal rights with respect to the marine and coastal environment. These must be given practical recognition pending confirmation and further clarification by the courts and negotiated claims settlements in light of the High Court's decision in *Mabo*.

6) There is a need to improve education and access to information amongst the regional population. An Advisory Committee will assist in the coordination and dissemination of research findings.

7) Torres Strait Islanders, as repositories of valuable local knowledge, should be the preferred candidates for training, management and field roles. Current challenges in relation to changing technology and other socio-economic influences will require the integration of this knowledge with western scientific knowledge through technical training.

What will MaSTS do?

- Establish locally-based institutional arrangements to develop and coordinate marine resource management and monitoring within the Torres Strait.

- Provide information relating to the natural environments of Torres Strait and advise on rational regional responses to global environmental change.

- Develop environmental planning and management guidelines, including provision of environmental impact appraisals to minimise the environmental costs of development.

- Provide practical solutions to present problems associated with locally generated pollution, including problems of inorganic or toxic household and industrial waste.

- Seek to reduce existing and potential threats of marine pollution or contamination at the regional scale (eg. Papua New Guinea mining operations and international shipping).

- Provide a forum through which the policies of the Torres Strait Islanders will be effectively represented in the management and utilisation of renewable resources.

- Facilitate implementation of indigenous Islander rights in relation to Torres Strait Treaty re-negotiation, fisheries agreements, 'native title', and possible management tools such as 'sea closures', in light of the *Mabo* decision.

- Reconcile technological and traditional Islander management concepts and customs in a culturally acceptable marine strategy.

- Coordinate and further develop research programs jointly by outside researchers and local people to ensure cultural appropriateness, local content and relevance.

- Further develop and encourage education and training programs to enable Torres Strait residents to play a greater role in marine management and marine-related employment.

MaSTS — Immediate Action Plan

- Discussion Paper, *Towards a Marine Strategy for Torres Strait (MaSTS)*, released for public comment
- Advisory Committee to MaSTS established
- Workshop on 'Marine Management Opportunities and Constraints: Options for Change' to be held on Thursday Island in mid June 1993
- Publication of Workshop Proceedings

MaSTS — Coordinating Team

MaSTS is a project of the Island Coordinating Council (ICC) of Torres Strait, assisted by the Australian National University North Australia Research Unit.

Copies of this publication are available from the North Australia Research Unit (NARU)

PO Box 41321, Casuarina, Darwin
Northern Territory 0811, Australia
Telephone: 089-27 5688
Facsimile: 089-45 0752

For further information contact the MaSTS Coordinator, Dr Monica Mulrennan, NARU

TOWARDS A MARINE STRATEGY FOR TORRES STRAIT

(MaSTS)

Prepared by Monica Mulrennan (MaSTS Coordinator), with assistance from Peter Jull, Marjorie Sullivan and Nicola Hanssen (NARU), in consultation with the Island Coordinating Council, Victor McGrath and other Torres Strait Islanders

A project funded by the Ocean Rescue 2000 Program of the Commonwealth of Australia

Published jointly by the Australian National University North Australia Research Unit and the Torres Strait Island Coordinating Council

First published in Australia in 1993 by the Australian National University North Australia Research Unit, and the Torres Strait Island Coordinating Council.

The views expressed in this document reflect those of the Island Coordinating Council and the MaSTS Team and do not necessarily represent those of the Ocean Rescue 2000 Program.

Inquiries should be directed to the publisher, Australian National University North Australia Research Unit and the ICC, PO Box 41321, Casuarina, (Darwin), Northern Territory 0811, Australia. Telephone: 089-27 5688 FAX: 089-45 0752.

Further information is available from the MaSTS Coordinator, Dr Monica Mulrennan, North Australia Research Unit, Darwin

National Library of Australia
Cataloguing-in-Publication entry:

Mulrennan, Monica, 1963–

Towards a marine strategy for Torres Strait (MaSTS)

Bibliography.
ISBN 0 7315 1545 5.

1. Coastal zone management — Queensland — Torres Strait Islands. 2. Environmental policy — Queensland — Torres Strait Islands. 3. Marine resources conservation — Queensland — Torres Strait Islands. 4. Torres Strait Islands (Qld.) — Economic conditions. I. Australian National University. North Australia Research Unit. II. Torres Strait Island Coordinating Council. III. Title.

333.916416099438

Printed in Darwin by the Northern Territory University Printing and Publishing Division

Printed on recycled paper

To the Reader

Torres Strait is a unique and bountiful environment. It is also the home of a culturally distinct society, the Torres Strait Islanders. The links between people and place are inextricable. Our culture and livelihoods, our past history and future prospects all depend on the well-being of our sea and of the islands, reefs, and shores which make up the region.

It is small wonder that Torres Strait Islanders feel a sense of responsibility about what happens in the Strait. We have paid the social, economic, and environmental costs of past boom and bust, and of greed, policy miscalculation, and decisions made far from our regions. Today, with fears of sea level rise, ocean tanker accidents, mines, ports and other developments coming to the waters flowing into the Strait and the marine systems of which the Strait is part, we know that we must act.

The development of MaSTS — the Marine Strategy for Torres Strait — is our response. It is an open public process through which the knowledge, needs and priorities of the people of Torres Strait will be integrated with the expertise of natural and social science disciplines, and the resources of state and federal levels of government, to provide for the effective management of the marine environment. Over the years there have been many and various actions and studies carried out in our region. These have provided much useful information, and in the development of MaSTS we have drawn on them as our principal non-Islander source. The time has now come, however, when an overall strategic view, with comprehensive environmental management guidelines, is needed.

Torres Strait Islanders are the first people to be affected by any problems in the Strait. In recognition of this fact, the Island Coordinating Council (ICC) has taken on the responsibility for developing MaSTS. We know from past experience that unless Torres Strait residents manage this work, the results will be unsatisfactory, ending in squabbles, theorising, and inappropriate decision-making by interests and agencies far removed from and little concerned with our daily lives. Torres Strait is our life. We cannot abdicate our responsibilities for it.

As we embark on the development of this Strategy we must keep in mind two major political and legal facts which govern our region. The first is the existence of Islander rights predating European settlement of Australia, rights which have at last been recognised in the High Court's decision of June 1992 in Mabo. A whole network of indigenous laws, rights and obligations, as well as management customs and eco-system perspectives, lie beyond the present scope of Australian and Queensland law. Rather than fearing these, they strengthen the scope for sound environmental protection and sustainable development.

Another matter is the Torres Strait Treaty. This complex Treaty contains provisions which are virtually unique in international relations. Some of these are positive, but in the daily workings of the Treaty we have found many problems. Some of the more sensational of these have reached Australian television viewers, while our people live daily with social, health and other quieter problems. The Treaty was in many respects a novelty. We should recognise that we have learned from it and make revisions on the basis of experience and new realities. Among those realities are surely the Mabo decision and Australia's rethinking of the constitutional status of indigenous peoples.

The following pages are ideas and suggestions. They are meant to promote further discussion and consultation, both within the Torres Strait and within expert and official communities outside the Strait. They will be refined and revised in this process. What is written here is flexible and living; it is not the last word, but it is a beginning.

Consultation and cooperation are essential to the success of MaSTS. With the best intentions in the world, and all the money, we could not solve Torres Strait issues without the goodwill of governments and other interests based outside our region. An essential part of MaSTS implementation will be the re-education, re-orientation, and re-shaping of public bodies in order to meet the challenges identified. Within our region, we must build bridges to non-Islanders who are part of the Torres Strait community. They too are an essential component of this project.

All around Australia the need to manage the marine environment is gaining recognition. Here in the Torres Strait the challenge is particularly great. As we begin our work on MaSTS, outside interest groups are watching us. They have heard of what we are doing, and they know it is a pilot project which may offer some solutions to an Australia-wide problem. We are proud to be taking a lead in this process.

In the preparation of MaSTS we must especially thank MaSTS Coordinator, Dr. Monica Mulrennan of the North Australia Research Unit, Australian National University, and the Assistant Coordinator, Victor McGrath of Thursday Island. I can tell every reader that the few pages here are only a very small part of the enormous amount of research, documentation and consultation which has been undertaken to bring us this far.

Getano Lui, Jr.
Chairman, Island Coordinating Council

ISLAND COORDINATING COUNCIL MEMBERS

Chairmen and Chairlady of the Torres Strait Islands including Bamaga and Seisia Communities

Getano Lui (Jnr)
ICC Chairman
Chairman of Yam Island

Joey Nona
Chairman of
Badu Island

Waiaka Jawai
Chairman of
Bamaga Community

Donald Banu
Chairman of
Boigu Island

Murray Lui
Chairman of Poruma
(Coconut) Island

Walter Lui
Deputy Chairman of
Erub (Darnley) Island

Phillip Bigi
Chairman of
Dauan Island

Henry Garnier
Chairman of Keriri
(Hammond) Island

Oza Bosun
Chairman of Kubin
Community, Moa Island

Daynee Mareko
Chairlady of
Mabuiag Island

Ron Day
Chairman of Mer
(Murray) Island

Terry Waia
Chairman of
Saibai Island

Joseph Elu
Chairman of
Seisia Community

Arthur Stephen
Chairman of Ugar
(Stephen) Island

Wilhelm Pedro
Chairman of St Pauls
Community, Moa Island

Allan Ketchell
Chairman of Tamwoy
Thursday Island

Nelson Billy
Chairman of Warraber
(Sue) Island

Joseph Mosby
Chairman of Masig
(Yorke) Island

Contents

List of abbreviations

AIDAB	Australian International Development Assistance Bureau
AIMS	Australian Institute of Marine Science
AMCS	Arctic Marine Conservation Strategy
AMSA	Australian Maritime Safety Authority
ANPWS	Australian National Parks and Wildlife Service
ATSIC	Aboriginal and Torres Strait Islander Commission (Commonwealth)
CDEP	Commonwealth Development Employment Projects
CEPA	Commonwealth Environment Protection Agency
CEPANCRM	Contract Employment Program for Aboriginals in Natural and Cultural Resource Management
CLC	Central Land Council
CMT	Customary Marine Tenure
CSIRO	Commonwealth Scientific and Industrial Research Organisation
DASET	Department of Arts, Sport, the Environment and Territories (Commonwealth)
DFAT	Department of Foreign Affairs and Trade
DOGIT	Deed of Grant in Trust
EIS	Environmental Impact Statement
EMC	Environmental Management Committee
ENSO	El Nino-Southern Oscillation
ESD	Ecologically Sustainable Development
GBRMPA	Great Barrier Reef Marine Park Authority
IBIS	Island Board of Industry and Service
ICC	Island Coordinating Council of Torres Strait
IDC	Interdepartmental Committee
IMO	International Maritime Organisation
IPCC	Intergovernmental Panel on Climate Change
ITCZ	Intertropical Convergence Zone
IUCN	International Union for the Conservation of Nature and Natural Resources
MARPOL	Marine Pollution Act (International Convention for the Prevention of Pollution from Ships 1973/78)
MaSTS	Marine Strategy for Torres Strait
NARU	North Australia Research Unit
NLC	Northern Land Council
OR2000	Ocean Rescue 2000
OSSM	On Scene Spill Model

PCDS	Priority Community Development Strategy for Torres Strait
PNG	Papua New Guinea
PZJA	Protected Zone Joint Authority (or TSPZJA)
RAC	Resource Assessment Commission
SOMER	State of the Marine Environment Report
SPREP	South Pacific Regional Environment Programme (Convention on the Conservation of Nature in the South Pacific)
TSMCS	Torres Strait Marine Conservation Strategy
TSPZ	Torres Strait Protected Zone
UN	United Nations
UNCED	UN Conference on Environment and Development
UNCLOS	UN Convention on the Law of the Sea
UNEP	UN Environmental Programme
WWF	World Wildlife Fund for Nature

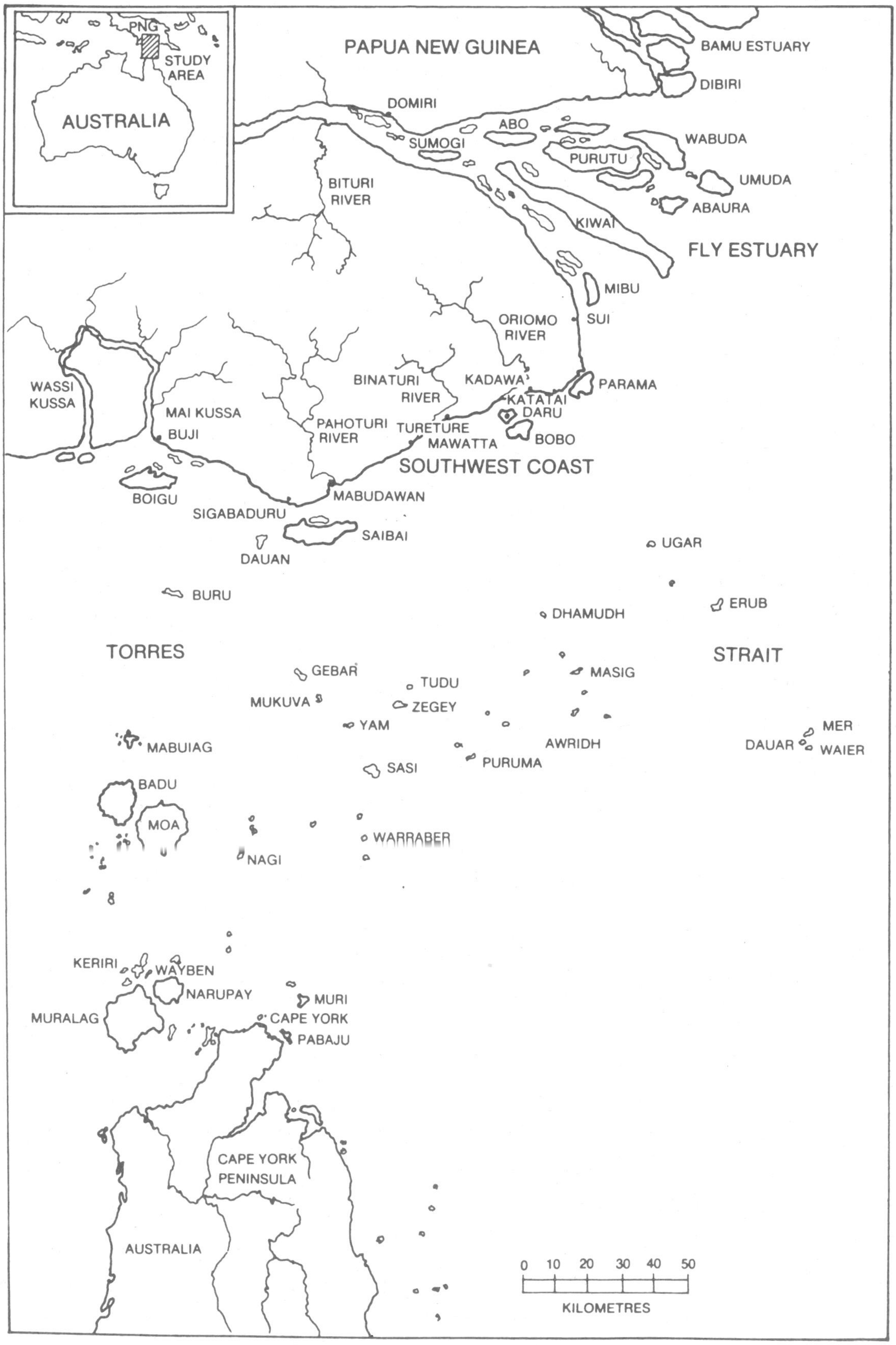

Figure 1. Map of the Torres Strait region
Source: Lawrence & Cansfield-Smith 1991, 1

Introduction

1

1.1 The need for MaSTS

The relationship between indigenous peoples and the environment has become the fundamental defining characteristic of indigenous identity. Unless practical environmental action is taken soon, however, this association will become an empty one in many places. Torres Strait is one such region under threat.

The Torres Strait Islanders still live in their traditional homeland, despite the removal of some of their people from certain islands during the period of European settlement. (Others have moved more recently to the mainland for economic reasons and because of island flooding.) Islanders are still largely dependent on the food and livelihoods provided by the sea. A variety of threats to the marine environment have emerged in recent years which have received public recognition, eg, potential contaminants from the Ok Tedi Mining Operation in neighbouring Papua New Guinea (PNG), as well as the considerable attention of specialists (Lawrence & Cansfield-Smith 1991). For these reasons Islanders strongly support ecosystem management and ecologically sustainable development. In 1991 they issued a telling statement of their concerns and needs, 'Principles and Objectives for the Future of Torres Strait'. It was here that the Marine Strategy for Torres Strait (MaSTS) was born.

> A comprehensive conservation and sustainable development strategy is needed as the foundation for future public policy and economic development in the Torres Strait region. A Marine Strategy for Torres Strait (MaSTS) would be the largest element in this approach, building on the unique relationship between Islanders and the sea. This relationship is as important for cultural traditions and social forms as for our livelihoods and future well-being (Principles and Objectives for the Future of Torres Strait, May 16 1991, 10, corrected version).

The Brundtland Report, which has done so much to set the international agenda on environment and development issues in recent years, is very clear on the need for the countries of the world to support and 'empower' indigenous peoples in those remaining homelands where they have survived for thousands of years (Brundtland *et al* 1987, 114–16).

Importantly, it links environment and resource management with policies for social justice, a link too often forgotten by governments:

> The starting point for a just and humane policy for [indigenous] groups is the recognition and protection of their traditional rights to land [and sea] and the other resources that sustain their way of life — rights they may define in terms that do not fit into standard legal systems. These groups' own institutions to regulate rights and obligations are crucial for maintaining the harmony with nature and the environmental awareness characteristic of the traditional way of life. Hence the recognition of traditional rights must go hand in hand with measures to protect the legal institutions that enforce responsibility in resource use. And this recognition must also give local communities a decisive voice in the decisions about resource use in their area (Brundtland *et al* 1987, 115–16).

The Report also recognises the larger national and international context:

> These communities are the repositories of vast accumulations of traditional knowledge and experience that links humanity with its ancient origins. Their disappearance is a loss for the larger society, which could learn a great deal from their traditional skills in sustainably managing very complex ecological systems. It is a terrible irony that as formal development reaches more deeply into rain forests, deserts, and other isolated environments, it tends to destroy the only cultures that have proved able to thrive in these environments (Brundtland *et al* 1987, 114–15).

Indeed, those wide interests seem to be the main reason advanced by the Brundtland Report for 'Empowering Vulnerable Groups' (as the Report heads its section on indigenous peoples).

The wider national purpose for action on marine issues in Australia has been focussed by a number of recent reports and enquiries. Programs and policies directed from the top have been too often considered the answer to all ills. Such approaches are rarely appropriate for distinct cultural groups like Torres Strait Islanders or for sensitivity to regional uniqueness. More recently, some efforts have emerged to 'ensure that the interests of Aboriginal and Torres Strait Islander people are reflected in the management of [a particular] Area' (GBRMPA 1992). Nevertheless there persists a sense that indigenous peoples must somehow be directed from above in order to achieve good management.

Australia has now signed Agenda 21 of the Rio Earth Summit. This substantial document provides clear guidelines for implementing the Brundtland approaches cited above in indigenous homelands such as the Torres Strait. A thorough change in approach is needed in Australia. MaSTS is an attempt by the Island Coordinating Council of Torres Strait to carry out such a change — for the good of the people of the Torres Strait and for all Australians who care about the marine environment.

1.2 The Torres Strait region

The Torres Strait is a 150 km wide passage between Cape York Peninsula of north-east Australia and the south-west coast of PNG. It incorporates the northernmost part of an internationally significant ecosystem, the Great Barrier Reef, other extensive reef areas, islands, islets, cays and mangroves, as well as some of the most extensive sea-grass areas in the world.

The region is culturally unique within Australia, being home to Australia's indigenous Melanesian people — the Torres Strait Islanders. Today there are an estimated 6,300 Islanders living on 16 islands within the Strait and two additional communities on the northern tip of Cape York. Approximately 15,000 now live outside the Torres Strait region on the Australian mainland, although there are indications of some return migration (Arthur 1990).

The Islanders have a strong seafaring and trading tradition and hence a close relationship with the tropical seas, coast and regenerating reefs of the region. The significance of the sea as the basis of their livelihoods and food sources can be appreciated by the fact that average rates of consumption of seafood in the Torres Strait are among the highest in the world (Johannes & MacFarlane 1991). The Islanders now wish to create and sustain a self-supporting economy based on their marine resources.

In recent years the Torres Strait has seen a substantial growth in tertiary (service) sector jobs, although these jobs have often gone to outsiders. The proliferation of public sector bodies, both service delivery and policy planning functions, has been welcomed for the increased attention it has brought to local problems. Lack of coordination and lack of decision-making accountable to local elected bodies has brought new problems, however.

The Torres Strait community has significant social and economic problems in the form of high unemployment and under-employment, limited infrastructure, acute health and education needs, and alcohol abuse and related problems. The lack of job prospects in particular, coupled with feared loss of accustomed social and cultural traditions and stability, have generated a corrosive sense of alienation among many people. The great post-war out-migration of Islanders may have eased some pressures, but it has also highlighted the regional disparity experienced between this corner of Queensland and the rest of Australia.

The major determinants of federal and state government policy in Torres Strait today are inevitably the difficulties of maintaining Australia's territorial integrity and a national frontier with countries to the north; protecting and managing a sensitive and hazardous physical marine environment which includes important international shipping routes; providing equitable living conditions and opportunities to a remote and chronically depressed region; and fulfilling evolving international standards for the survival, cultural autonomy, and appropriate political institutions of a distinct indigenous people. This is the policy and political context in which MaSTS must exist no matter what changes of party or personality may occur in Canberra, Brisbane or the Torres Strait.

1.3 Purpose of MaSTS

The purpose of the Marine Strategy for Torres Strait (MaSTS) was defined at a first consensus-building workshop in September 1992:

> to establish a comprehensive management framework which will permit optimum exploitation of the region's limited resource base, consistent with the needs of indigenous Torres Strait Islanders, sustainable development and minimal environmental disturbance (ICC & MaSTS team).

Agreement was also reached at this workshop to change the proposed title of the Strategy from the Torres Strait Marine Conservation Strategy (TSMCS) to the Marine Strategy for Torres Strait (MaSTS). The new title was thought to embrace more firmly the need to ensure the health and well-being of the marine environment *and* to provide for the sustained utilisation of its marine resources.

The Torres Strait Regional Council (ATSIC) has recently developed a Regional Development Plan to address the economic, social and cultural issues of the Torres Strait. MaSTS will reflect these socio-economic and cultural needs, and provide an essential environmental perspective to the Regional Plan. MaSTS is designed as a prototype marine and coastal management plan enabling Torres Strait Islanders and other local residents to work more effectively with Queensland and Commonwealth governments. It is intended to locate decision-making, management and accountability in the Torres Strait Region to the greatest extent possible. It is also anticipated that MaSTS will upgrade resource planning and environmental management in Torres Strait.

Over recent months discussion and research have been building up to the development of MaSTS. Numerous Australian and overseas officials, academics, indigenous groups and individuals have recognised that MaSTS will be a significant and promising step: in the practical implementation of ecologically sustainable development (ESD); in the search for comprehensive Australian coastal zone management; in 'grass roots' environmental action and administration; and as a positive approach to the socio-economic problems of indigenous peoples. However, the most important statement of purpose has already been quoted above from the ICC's 'Principles and Objectives':

> A comprehensive conservation and sustainable development strategy is needed as the foundation for future public policy and economic development in the Torres Strait region (ICC 1991).

1.4 Principles of MaSTS

MaSTS is guided by the following principles, both in its development and implementation phases:

1) The natural resources of the Torres Strait are not unlimited: there are limits to the marine environment as a pollution sink and as a provider of renewable resources. There is, therefore, a need to protect the marine environment and its resources for present and future generations.

2) Effective resource and ecosystem management require that local residents play a leading role in decision-making processes. Many current policies and programs involving the management, protection and utilisation of the Torres Strait marine environment will need to be amended to provide for the greater involvement and representation of Torres Strait Islanders.

3) Many current policies and administrative practices are characterised by complex and overlapping jurisdictional responsibilities. There is a need to coordinate and rationalise existing arrangements and develop a comprehensive regional approach to the management of the Torres Strait marine environment.

4) The Torres Strait economy is based in large part on natural renewable marine resources. The needs of the Torres Strait

Islanders, as the principal users and dependants on the marine environment, must receive priority in any competition for these resources.

5) Torres Strait Islanders have legal rights with respect to the marine and coastal environment. These must be given practical recognition pending confirmation and further clarification by the courts and negotiated claims settlements in light of the High Court's decision in *Mabo*.

6) There is a need to improve education and access to information amongst the regional population. An Advisory Committee will assist in the coordination and dissemination of research findings.

7) Torres Strait Islanders, as repositories of valuable local knowledge, should be the preferred candidates for training, management and field roles. Current challenges in relation to changing technology and other socio-economic influences will require the integration of this knowledge with western scientific knowledge through technical training.

1.5 The Strategy in operation

The Marine Strategy for Torres Strait (MaSTS) is intended to provide **an overall framework** for the conservation, sustainable utilisation and management of the Strait's marine ecosystems.

At this stage MaSTS provides only provisional policy objectives and principles which reflect the needs and concerns of the Islanders. In its next stage MaSTS will explore opportunities and options for change in existing arrangements.

The Strategy has no legislative basis at present. However, its objectives and policies are already partly enshrined within the international requirements undertaken by Australia as part of the Torres Strait Treaty — requirements which have yet to be fully implemented (see Section 2.1).

The success of MaSTS is ultimately dependent on an ongoing committment from the Australian government to provide the support and resources which will ensure its effective implementation. This will include a financial committment and the provision of a variety of regulatory and administrative, and even legislative, action by various levels of public authority.

1.6 Geographic extent of MaSTS

The Marine Strategy for Torres Strait (MaSTS) applies to all marine waters bounded in the west by the Carpentaria Shoal near longitude 141° E and in the east by the edge of the Great Barrier Reef along longitude 144° 30' E. To the south the extent of MaSTS is delimited by latitude 10° 41' S to the east of the tip of Cape York and to the west of Cape York by latitude 11° S, including the coastal strip between the tip of Cape York to a point south of Crab Island. The northern boundary extends eastward from the Irian Jaya/PNG border, along the PNG coast to Parama where it diverges from the coastline along latitude 9° S (see Figure 1).

The Strategy deals explicitly with marine ecosystems and associated resources but also applies to the islands, cays and reefs surrounded by these waters. Coastal environments of the northern tip of Cape York are also included as these areas are relevant to the predominantly Islander communities at Bamaga and Seisia. The Aboriginal communities of Umagico, Mapoon and especially Injinoo, as the original inhabitants of the area, may also wish to play a role in the Strategy.

The inclusion of coastal environments within the Strategy is important because they impinge on the quality of the marine environment. Similarly, because MaSTS deals with the quality of the marine environment, its interest extends beyond the immediate area of the Strait to any waters discharging into or flowing through the region.

1.7 MaSTS and other conservation strategies

Much of the groundwork already exists for a Marine Strategy for Torres Strait. Article 13 of the Torres Strait Treaty between Australia and PNG, requires both parties *to take legislative and other measures to protect and preserve the marine environment in and in the vicinity of the Protected Zone.*

PNG is currently carrying out a pre-feasibility study to develop an environmental management plan as the basis for its responsibility in the Strait. Effective liaison between those involved in the formulation of the PNG plan and MaSTS will be essential as many activities undertaken by PNG and Australia ultimately affect the marine resources of the Torres Strait. This will require more coordination.

Various Commonwealth initiatives such as the Torres Strait Baseline Study, the pioneering Ecologically Sustainable

Development Reports, and economic studies from the 1988 federal Interdepartmental Committee (IDC) Report on the Torres Strait will provide both philosophical and specific underpinnings for MaSTS.

The latest version of the World Conservation Strategy, *Caring for the Earth: a strategy for sustainable living* (Earthscan 1991), set a helpful pattern for planning the sustained use of living resources including a number of specific priority actions for oceans and coastal areas. Australia has also developed its own National Conservation Strategy and is currently preparing a National Marine Conservation Strategy, through the Ocean Rescue 2000 Program. MaSTS is seen as 'a regional pilot project within this strategy which, if successful, could be used as model for other community projects elsewhere in Australia' (ICC correspondence from DASET, 1992).

Other federal initiatives include the proposed National Coastal Zone Management Strategy, and the Coastal Zone Inquiry of the Resource Assessment Commission which is examining Aboriginal and Torres Strait Islander management arrangements and issues as part of its work (Smyth 1992). (ICC and the MaSTS team have both contributed considerably to this RAC project.)

The Great Barrier Reef Marine Park shares a common boundary with the Torres Strait. This World Heritage area has been gazetted as a multiple use marine park, under the management of the Great Barrier Reef Marine Park Authority (GBRMPA). The broad management aims for this extensive marine area include the ecologically sustainable management of the area, the maintenance of biodiversity and marine ecosystem values, continued commercial, recreational and conservation issues of the area, and flexible consultative management involving a wide sector of the Australian public, both indigenous and non-indigenous. The management of the Torres Strait marine system has a direct influence on the Marine Park. Actions taken in the Park impact directly or indirectly on the Torres Strait. Given the connectivity and linkages in marine ecosystems and environments, consultation and cooperation between GBRMPA and ICC are essential to ensure a holistic and integrated approach to marine environment management.

At the state level, MaSTS will draw upon many Queensland initiatives and dovetail with them. The proposed Queensland Conservation Strategy, the Coastal Protection Strategy, and the Cape York Peninsula Land Use Study are examples. The essential difference between these strategies and MaSTS is that the latter has the perspective of the regional population of Torres Strait at its centre rather than as one interest among many.

The development of MaSTS has more in common with initiatives from overseas, for instance Canada where three federal departments (Fisheries and Oceans; Environment; and Indian Affairs and Northern Development) plus two northern territory governments represented by their Renewable Resources departments, and Inuit representative bodies have designed whole new approaches to comprehensive marine environmental management. The Arctic Marine Conservation Strategy (AMCS) provided some useful ideas for MaSTS, and the Arctic Environmental Strategy, which later subsumed AMCS, highlights the comprehensive 'ecosystem' approach to maintaining the integrity of the natural environment while centrally involving the indigenous population. The MaSTS project is also in touch with similar work going on in Norway, Alaska and the South Pacific, and has been sharing information and ideas back and forth.

1.8 Agencies involved in strategy development

MaSTS is being developed under the Ocean Rescue 2000 Program (OR2000) established by the Commonwealth Department of Arts, Sport, the Environment and Territories (DASET). This 10-year program is aimed at the conservation and sustainable use of marine resources. It includes five key elements: an Australian Marine Conservation Strategy; a State of the Marine Environment Report (SOMER); a National Representative System of Marine Protected Areas; a National Marine Education Program; and a National Marine Information System. Within the Commonwealth government, administration of the OR2000 program is overseen by a Steering Committee comprising the Secretary of DASET, the Chief Executive Officer of Australian National Parks and Wildlife Service (ANPWS), and the Chairman of the Great Barrier Reef Marine Park Authority (GBRMPA).

The Island Coordinating Council (ICC) of Torres Strait has initiated and is leading the development of MaSTS. As lead agency, the ICC is involved in an extensive consultation process with interested parties including Islander people, industry, federal and state government departments, educational institutions and non-government organisations. Workshops, providing for more general representation from the Islanders, ensure consensus on the purpose and principles which form the basis of the Strategy.

Victor McGrath of Thursday Island is Assistant Coordinator to MaSTS. An Advisory Committee consisting of representatives from these various interest groups will be established during the early stages of strategy development and will hold regular meetings.

The North Australia Research Unit (NARU) of the Australian National University, under the directorship of Professor David Lea, is assisting ICC in the development of MaSTS. Dr Monica Mulrennan from NARU is the Coordinator of MaSTS, with assistance from Mr Peter Jull on political and policy issues and Dr Marjorie Sullivan on environmental planning and sustainable development. Ms Nicola Hanssen is a research assistant to MaSTS.

Jurisdiction, rights and institutional framework

2

In recent months the Torres Strait Islanders through the ICC have engaged in consensus building workshops and discussions to develop MaSTS. Many of the issues and concerns raised during those discussions are outlined below. Some of these issues and concerns are beyond the scope of the Strategy — MaSTS is specifically concerned with management, protection and sustainable utilisation of the Torres Strait marine environment. Nevertheless some of those further issues and concerns provide an important context and background to the Strategy.

2.1 Torres Strait Treaty

The Torres Strait Treaty is a bilateral arrangement between the governments of Australia and PNG. It was signed in December 1978 after six years of negotiation and its enabling legislation, the Commonwealth *Torres Strait Fisheries Act 1984*, was passed by the Australian Parliament in April 1984. It was ratified and entered into force on 15 February 1985.

The Treaty is concerned with sovereignty and maritime boundaries, protection of the way of life and livelihoods of traditional inhabitants, and protection of the marine environment.

Contrary to its initial expectations, the Torres Strait Treaty is now identified by the ICC as the most significant jurisdictional/ administrative problem with which it has to deal.

> The big problem at the moment is with the Torres Strait Treaty ... at one stage when it was first implemented we all thought that it was there to protect our rights but now it seems that it is there to protect everyone else except us — the Torres Strait Islanders themselves — and this problem has built up over the years in frustrations (Getano Lui, quoted in Jull, Mulrennan & Sullivan 1993).

Another problem reported by Islanders is that the Treaty has created a logistical and bureaucratic nightmare (Jull, Mulrennan & Sullivan 1993). Simple community-based tasks such as policing, seeking funding, and exploring new enterprise initiatives, are now plagued with paperwork, while needed resources disappear into administrative costs.

Outside experts have considered the Treaty to be good in principle with potential to become a powerful tool (Laffan 1991). They acknowledge, however, that it has proved to be disappointing in many respects.

> For a document replete with promises and good intentions, it has seemed to lack teeth! (Allen 1993)

2.1.1 The Protected Zone

One of the special provisions of the Treaty is the Torres Strait Protected Zone (TSPZ), an area encompassing most, but not all, of the inhabited islands in the Torres Strait region (see Figure 2). The purpose of the TSPZ, as described in Article 10, is:

> to acknowledge and protect the traditional way of life and livelihood of the traditional inhabitants including their traditional fishing and free movement.
>
> to protect and preserve the marine environment and indigenous fauna and flora in the vicinity of the Protected Zone.

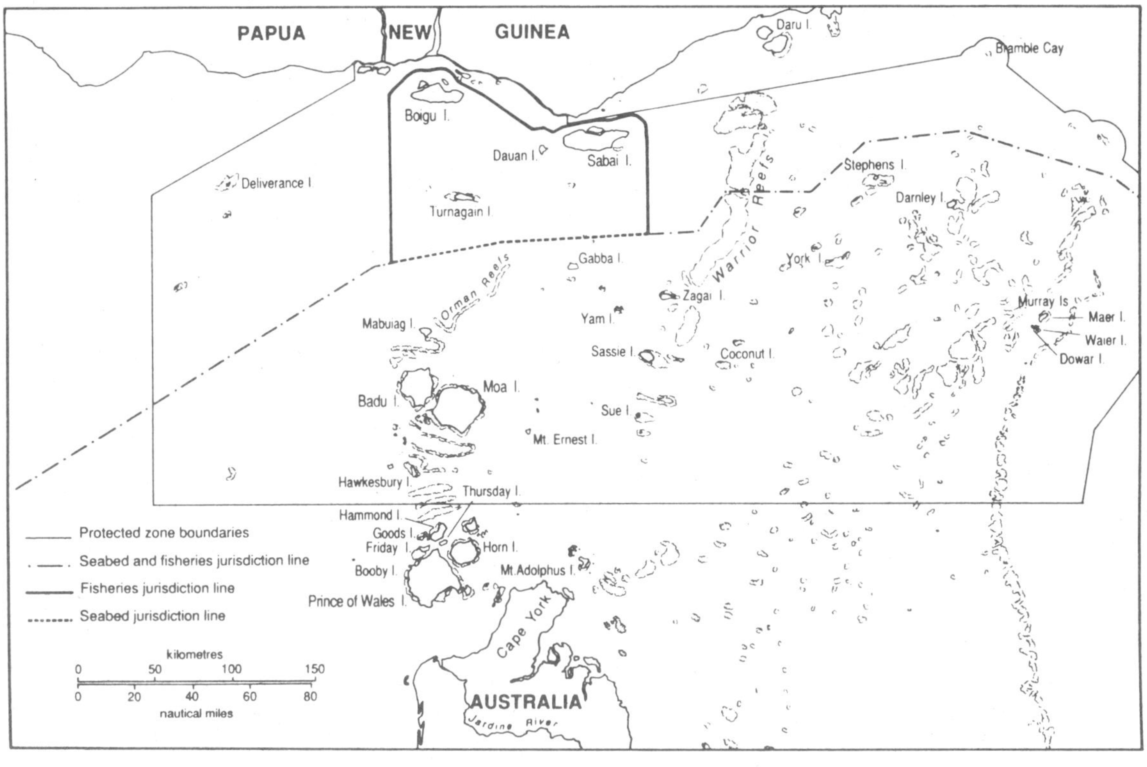

Figure 2. Lines of jurisdiction agreed under the Torres Strait Treaty
Source: Babbage 1990, 34

Free movement and *traditional activities* including *traditional fishing* are defined in Article 1 and further clarified in Article 11. *Traditional* is to be *'interpreted liberally and in the light of prevailing custom'*. A recurring difficulty is that what Islanders agree is the prevailing custom, may clash with governmental fisheries specifications, particularly in relation to commercial equipment (Murray 1992).

Many of the problems reported seem due to the disparities between the Australia and PNG economy with the result that PNG *traditional* visitors have more to gain from *free movement* in the Torres Strait region than their Australian counterparts. The international border has also become a gateway for other nationalities seeking entrance to Australia. Problems arising from this free movement provision include: illegal fishing, directly or in support of others; illegal immigration; abuse of welfare facilities; drug-running; gun-running; and other customs and quarantine violations.

Islanders, out of a sense of regional responsibility, have been working through the ICC to obtain from both Queensland and federal governments a larger role in managing the border. A partial response in a recent Queensland government indigenous policy study, *Towards self-government* (Legislation Review Committee 1991), has been encouraging, but further action is needed.

2.1.2 Management in the Protected Zone

The Protected Zone Joint Authority (PZJA) was set up, under Section 30 of the *Torres Strait Fisheries Act 1984* (Cwlth), to manage the consultative and decision-making mechanisms of fisheries in the Protected Zone. It is responsible for licensing, control, protection and conservation of all fisheries, including turtle and dugong, in accordance with Commonwealth law. Queensland manages all other commercial and recreational fisheries in the Australian area of the Torres Strait.

The PZJA has developed a policy of increasing opportunities for Islander participation by preventing further expansion of non-Islander involvement in fisheries. No additional non-Islander participants are being licensed to fish in Joint Authority fisheries (Elmer & Coles 1991), and growth in the tropical rock lobster, mackerel, barramundi and pearl shell fisheries is now reserved for Torres Strait Islanders (TSPZJA 1990). These Islander-orientated initiatives are, however, offset by the sheer strength of commercial fishing interests in the area. Commercial fishing

interests are seen by Islanders as being overly represented on consultative committees set up by PZJA (Murray 1992).

'Community fishing' is the area in which the PZJA sees Islander participation in fishing increasing. This is defined in Section 3 of the *Torres Strait Fisheries Act 1984*, as 'commercial fishing carried out by traditional inhabitants'. Community based fishing industries are now operating on several islands where crayfish tails (tropical rock lobster) and mackeral are stored in freezers before being sold to interstate buyers. Profits are distributed between the Island Council and the participating fishermen.

In the past weeks the Islanders have been approached by several different agencies involved in the management of dugong, among other species. For example, CSIRO Fisheries invited Islander representatives to attend a one-day workshop on the results of their assessment of the marine catch from the Protected Zone. ANPWS also proposes to discuss the management of dugong at the forthcoming EMC meeting. Meanwhile the Islanders are also involved in discussions with the Queensland Department of Environment and Heritage regarding the proposed development of wildlife conservation plans, which will include dugong. There are several other initiatives also involved in dugong management.

From the Islanders viewpoint, the existing arrangements seem unnecessarily complicated. Greater coordination is needed to ensure that the Islanders are presented with a comprehensive package for dugong management rather than having to deal with the separate jurisidictions and processes of a number of separate agencies.

The procedures involved for Islanders applying for licences can be complicated. As Allen states, even when such problems are overcome,

> Until exclusive interests in the sea, the sea bed and their resources are established in law, indigenous people will remain dependent on the exercise of such privileges to fish as may be granted as a matter of noblesse oblige (Allen 1993).

A Torres Strait Fishermen's Association exists to look after the interests of local fishermen and there seems to be much that can be done to improve the situation initially by negotiations at a local level through this Association.

2.1.3 Protection and conservation in the Protected Zone

The Torres Strait Treaty provides the primary basis of control and protection of the Torres Strait marine environment. An Environmental Management Committee (EMC) provides the mechanism for communication while a Baseline Study in Australia and relevant studies in PNG, assessing the impact of mines in the upper Fly River catchment of PNG and the effects of their effluents on Torres Strait, provide information. A commitment by the federal government to this four-year Baseline Study (which commenced in 1989) raised high levels of interest and expectations in Torres Strait. The Study is, however, concerned specifically with the biological and ecological identification and effects of heavy metals and was never intended to embrace all of the environmental issues and concerns of the Torres Strait.

Article 13 of the Treaty explicitly provides for the protection of the Torres Strait marine environment. For example Section 2c:

> obliges each party to minimise to the fullest practicable extent pollution or other damage from installations and devices used in the exploration and exploitation of the natural resources of the seabed and subsoil thereof (Torres Strait Treaty, 13[2c]).

Section 5 relates specifically to the assessment of potential impacts of planned activities on the marine environment, while Section 6 provides for control over existing activities which already impact on the environment. But, as O'Rourke noted in the 1988 Report of the Interdepartmental Committee (IDC),

> Australia has no legislation in place to implement Article 13. Implementation depends on the goodwill of the Queensland government to require adequate environmental protection measures for developments and to consult the Commonwealth (O'Rourke 1988, 14).

The case of the Horn Island gold mine illustrates this point. There the Commonwealth had the opportunity to comment on guidelines for the EIS which the Queensland government required, but it was not consulted over the decision to grant a mining lease, nor over the final conditions for environmental protection which were imposed. The Commonwealth seems unable to guarantee the requirements of the Treaty at present. Some mechanism to implement the provisions for environmental management under Article 13 is urgently required (O'Rourke 1988).

Similar problems exist in relation to Article 15 of the Treaty which places a 10 year moratorium on any mining and/or drilling activity of the seabed within the Protected Zone. Article 15 suggests that when the 10 year moratorium ends in 1995 'the parties may agree to extend [that period]'. There are no guidelines as to what should occur if the parties agree not to extend the moratorium (Murray 1992, 29).

The Treaty is recognised as having a 'major environmental protection dimension' and that 'although originally drawn up in the 1970s... it is still a good tool for achieving environmental protection of the Torres Strait region' (Laffan 1991, 453). It contains many elements of sound environmental — particularly marine — protection and management and is sympathetic towards the significance of traditional activities in the region. The Treaty has not, however, lived up to its potential and MaSTS will need to address many issues relating to its environmental management provisions.

2.1.4 Islander consultation

Article 30 (Consultations) states that 'the parties shall consult, at the request of either, on any matters relating to this treaty'. Several committees and consultative mechanisms have been set up to assist in this regard. However, it seems that Torres Strait Islanders have no express abilities under the Treaty or the *Torres Strait Fisheries Act* to initiate consultations of any kind, except at a local level (Murray 1992). For example, international problems with regard to fishing must be dealt with by the Commonwealth government, as only it has the capacity to enter international agreements. For Torres Strait Islanders, advisory committees are useful to allow their voices to be heard, but the end of the line of action is a Foreign Affairs or Primary Industries Minister who may or may not feel it necessary to consult with PNG (Murray 1992). To the Islanders this seems a cumbersome and frustrating way to tackle immediate and local concerns.

An illustration of communication problems is provided by the recent concerns expressed by the ICC in relation to the potential risk of oil pollution associated with increasing tanker traffic to and from the Kutubu oil facility in PNG. According to the Queensland Coast and Torres Strait Pilot Service, of the 1,500 ships that transit the Strait each year some 200 are oil and chemical tanker transits. In April 1992 the ICC was informed that an estimated addition in traffic of one or two tankers per year would collect oil from the Kutubu oil terminal and pass westward through the Torres Strait. In May 1992 the ICC was advised that the previous estimate had been revised and that the Pacific Chevron Company now expected a maximum transit of one or two tankers per week. By June 1992 the ICC was again notified and advised that the terminal operator AMPOLEX had again amended its estimate of tanker traffic to approximately two tankers per month. A total of 24 tankers were reported at the recent JAC meeting (November 1992) to have passed through the Strait over a two month period.

It is hardly surprising that present arrangements generate disaffection, confusion, frustration, and ill-will among the traditional inhabitants of the region.

Decisions affecting the regional environment are generally outside the powers of the Islanders. The extent of government consultation on such matters is in many cases little more than receipt of official notification of decisions already taken by government bodies in Canberra or Brisbane. Such unworkable arrangements corrode the moral authority of the very systems they are intended to maintain.

The recent case of the Taiwanese fishing vessel, the Sheng Fu No. 16, provides an illustration of the extent of Islander involvement in local decision making. The vessel was wrecked on Turu Cay, approximately 60 nautical miles north-west of Thursday Island. Efforts to retrieve three nets belonging to the boat, estimated to be each 2–3 nautical miles in length, were unsuccessful. These loose gill-nets were left to drift through the waters of the Strait, trapping and killing fish and other marine life. The wreck also held (illegally) two monkeys, two dead dugong and rotting fish carcasses amounting to more than the total legal catch for boats in the Torres Strait over an entire year (ICC Memo, 2 June 1992).

Regarding removal of the vessel, the Australian Maritime Safety Authority (AMSA) was advised in June 1992 by Queensland National Parks and Wildlife Service that 'the cay was not of significant environmental interest to warrant removal'. This advice, together with consideration of the fuel type, led AMSA to conclude that 'there was no significant threat to the marine environment'. This decision was confirmed in early August 1992 by the Department of Shipping and Aviation in a letter informing the ICC that as the Commonwealth Environment Protection Agency (CEPA) did not consider that the vessel constituted an environmental threat, a decision had been taken to leave the wreck on the cay. By mid August 1992 CEPA had modified this decision advising that removal of the 130–150 tons of diesel oil was necessary and that the vessel should then be left in a stable and environmentally non-threatening condition.

The fact that Turu Cay lies within a highly sensitive environment of significance to the Islanders — including the sole recognised dugong sanctuary, a major Flatback turtle rookery, and important pearl shell grounds — is only one of the issues here. The extent and nature of their involvement in decisions relating to such matters and the fact that the question of responsiblility to investigate the problem was disputed for some time between the Department of Foreign Affairs and Trade and the Australian Maritime Safety Authority is perhaps of more fundamental concern.

2.2 International support

One of the main issues discussed at the Marine Workshop of the Conference, *Surviving Columbus: the indigenous environment today and tomorrow* (co-sponsored by NLC, ICC, CLC and NARU, Darwin, October 1992), was the existence of international treaties and conventions through which the Islanders could

ensure that the two parties involved in the Treaty uphold their responsibilities and the premises of the Treaty.

Many provisions contained within UNCLOS, the *United Nations Convention on the Law of the Sea*, are particularly relevant to the Torres Strait. For example, Part IX contains a Deep Seabed Mining Regime, while Article 194 places an obligation upon states to act individually or jointly in order to prevent, reduce, or control pollution of the marine environment from any source. It has also been suggested that certain boundary problems associated with the Treaty could be resolved through UNCLOS (see Babbage 1990, 44). Although many nations have signed and ratified UNCLOS, their number has not yet been sufficient to bring it into force (Australia has yet to ratify UNCLOS).

A number of other global instruments, to which Australia and/ or PNG are signatories, deal either with marine pollution or conservation of living marine resources. These include the London Dumping Convention, International Maritime Organisation (IMO) Conventions controlling pollution from ships, and the Convention on the Conservation of Nature in the South Pacific (commonly referred to as the SPREP Convention). However, PNG does not have the capacity of Australia to execute its Treaty or international obligations in surveillance or investigation of its territorial waters. Recent events suggest that even the more affluent Australia lacks the means to carry out the surveillance necessary to effectively manage the Torres Strait border region.

A recent example of the failure of international convention to enforce regulation in the region was provided by the PNG Deputy Prime Minister and Fisheries Minister who in April 1992 issued a domestic gill-net licence to a Taiwanese-registered fishing vessel using nets which contravened the Wellington Convention — an international agreement (ratified by PNG) prohibiting fishing with long drift-nets in the South Pacific.

2.3 Local government

The Island Coordinating Council (ICC) was established under the *Queensland Community Services (Torres Strait) Act 1984*. This Act requires the ICC 'to secure the progress, development and well being of Islanders by way of Commonwealth or State government grants, and to employ such agents and servants as are necessary for the proper and efficient discharge of its functions and powers'.

The ICC is made up of the elected chairpersons of the individual Island Councils. Apart from the islands under the control of the State of Queensland Torres Shire (Thursday, Horn and Prince of Wales Islands) — and of course the Murray Islands — all other inhabited islands in the Torres Strait region and the two Cape York communities (Bamaga and Seisia) are controlled by Island Councils. Each Island Council holds title of its land under the Deed of Grant in Trust formula (DOGIT), governed by the *Queensland Community Services (Torres Strait) Act 1984*. DOGIT represents a form of inalienable title, that is, owners cannot sell it or give it away and it is revocable only by Act of Parliament (Brennan 1992). Avenues for exploration have, however, been opened by the precedent set by the *Mabo* judgement.

Despite the relative lack of funds and powers available to ICC, it enjoys a semi-autonomous position and has become the primary focus for Islander political, economic, social, cultural and environmental aspirations (Lui, UN Statement 1992).

> It was our 1988 call to secede from Australia that really was the turning point (Lui, quoted in Jull *et al* 1993).

Their demand for independence was a culmination of historic events, combined with a growing dissatisfaction with government policy and administration at all levels (Babbage 1990). It prompted the Commonwealth Department of Aboriginal Affairs to investigate and address the Islanders' listed concerns. A Report of the Interdepartmental Committee (IDC) on the Torres Strait Islands ensued, concluding that:

> Islander problems have been caused by a lack of coordination of the activities of all the agencies at the various levels of government, rather than the lack of funds. We found no evidence of an overall plan for regional development and, in particular, little sign of any coordinated efforts to plan for economic development of the area. In our opinion, many of the concerns of the Islanders could be met by the development of a more coordinated overall approach by governments (O'Rourke 1988, 6).

The IDC Report (1988) saw the instigation of ATSIC (the Commonwealth Aboriginal and Torres Strait Islander Commission) as a solution for improving government coordination in the Torres Strait region. The ATSIC Torres Strait Regional Council was established to provide a mechanism for Islander input into federal policy development in the region, while the ICC (established under State legislation) was to undertake a similar role in relation to State government programs.

The ICC has coordinated and developed many initiatives stemming from the IDC Report. For example, its search for a means of increasing local coordination and developing a local management structure resulted in the drawing up of a 5 year Regional Development Plan (see Arthur 1990 and Lea *et al* 1990). The ICC has gained considerable experience over recent years in identifying the specific needs of its people and in seeking solutions through contemporary governmental processes. To date, however, both State and Commonwealth governments have tended to want to administer the islands and Islanders through programs, rather than turning over management and decision-making to the Islanders themselves. This reluctance to allow indigenous people to run their own lives results in the continuation of many social problems and grievances.

'Thursday Island presents a case study of unplanned development by government agencies and lack of consultation with the local governing authority... or with Torres Strait Islander people' (O'Rourke 1988, 37).

In 1987 fifteen Commonwealth and eleven State departments were operating on Thursday Island (Kehoe-Forutan 1991). A more recent source suggests that some thirty-five public service departments are presently based on Thursday Island and servicing a population of 10,000 (Jull et al 1993). As a result, the roles of most of these organisations either overlap or lack focus and impinge upon each others ability to achieve concessions and goals. There is thus an urgent need for some type of localised coordination and/or management structure (Kehoe-Forutan 1991, 424).

On May 16, 1991, the ICC served notice in a letter and attachments to the Queensland Premier, its *Principles and objectives for the future of Torres Strait*. These principles and objectives represent a statement of Islander aspirations and needs, including socio-economic needs as well as environmental issues. The issues are wide ranging but the message is simple — there is a need to increase local control. This approach is not out of step, indeed it is entirely in keeping with Australian political and constitutional values. Norfolk Island is no threat to Australian unity, nor is the increasing Pitjantjatjara management of their territories detracting from national well-being.

Torres Strait is a unique and discrete geographical region, containing a variety of local island identities and cultures. It would make sense to devise a political and adminstrative system suited specifically to those needs.

Australia could save itself much trouble by learning from the experience of other countries with indigenous policy, notably other Anglo-American democracies, and from the leading Nordic countries in this field, Denmark and Norway. Land claims negotiations in northern Canada have centred on provisions for land, freshwater and marine management and new statutory decision-making bodies. 'Home rule' for Greenland's indigenous people includes similar environment and planning issues. Those countries learned by trial and error, but now share ideas, both between their indigenous peoples and their governments. Difficult issues of resource use, traditional ownership, development rights, and revenue shares have been central to all those situations. They are by no means impossible to solve.

2.4 Regional development

A comprehensive review of the economic structure of the Torres Strait is presented by Lea *et al* (1990). A general outline is provided below.

Torres Strait fisheries are estimated to be worth some $20–$30 million per year, the prawn and crayfish (tropical rock lobster) fisheries being the most valuable. Despite the significance of fisheries in the Torres Strait, few local people are employed by the big fishing companies operating in the region (Lea *et al* 1990). The majority of Islanders who are involved are part time and for them commercial fishing is seldom their only, or even primary, source of income (Arthur 1991). Furthermore, as mainland fishers hire almost all their employees and obtain most of their provisions from outside the region they probably contribute less to the local economy, despite an annual turnover of $14 million, than did the Horn Island gold mine when it was operating.

Community Development Employment Projects (CDEP) wages make up 42% of the total amount of outer Islanders' income — the merits and drawbacks of this dependency are discussed elsewhere (Arthur 1991). 'Easy' money is considered by many to lead to disincentive but the issue is more complex than that. Islander involvement in fisheries depends on their access to freezers and other facilities as well as to expertise. Johannes and MacFarlane (1991) believe that the major impediment has been the failure of education in the Torres Strait to prepare Islanders to fully exploit the opportunities provided by local marine resources. Both Arthur (1990) and Lea *et al* (1990) proposed some form of development agency to facilitate increased Islander involvement.

One suggestion was the establishment of a structure similar to that provided by the *Maori Fisheries Act 1989* in New Zealand which would allow government to grant all or part of the fisheries quota to the Islanders who would then be required to lease this out to Islanders and non-Islanders (Arthur 1991). Another proposal was to utilise the *Aboriginal and Torres Strait Islander Commission Act 1989* as a means of establishing an Islander Corporation which would assist and coordinate Islander participation in future 'community fishing'.

Arthur's (1990) assessment of potential options for economic development indicate, however, that the long-term prospects for increasing involvement in fisheries are not encouraging.

> in the longer term sustainable development in the Strait may be predicated on broadening the economic base by diversifying into other industries, rather than by simply expanding the local fisheries to unsustainable levels (Arthur 1990, 417).

The potential for local market gardening is partly limited by quarantine restrictions on inter-island movement of certain fruit and vegetables in the Torres Strait Protected Zone. All fruit and vegetables sold by the Island Board of Industry and Service (IBIS), a non-profit regional organisation with retail stores on 15/21 islands, are imported from mainland Australia. Opportunities need to be explored so that locally grown foodstuffs can supplement if not replace imported goods.

The potential for tourism development is generally dependent on the ease and cost of access, scenic beauty, activity opportunities, and cultural attractions. The cost of air transport to and between the islands of the Torres Strait is a major limiting factor especially when compared with the Great Barrier Reef islands which have similar natural attractions. Activities such as snorkelling, fishing, bird watching and diving have much appeal but are not unique to the region and the winter tourist season coincides with a time of rough and dangerous seas in the Strait. An additional constraint is that many Islanders would prefer tourism developments to take place at some distance from their community village or on uninhabited islands (Arthur 1991). Such developments are likely to exacerbate existing problems with infrastructure and services, particularly in relation to potable water supplies, electrical power and modern plumbing (Mulrennan 1992).

Despite these limitations and constraints, the Islanders believe that it is only a matter of time before outside investors move in (*Torres News*, 5–11 July 1991). The Aboriginal Development Commission identified three strategies for tourism development: exclusive high cost/high standard tourist complexes; medium standard activity-oriented developments; and wildlife-based developments (O'Rourke 1988). Several projects are presently under consideration. Their potential impacts will need careful assessment.

With regard to extractive industries, if the moratorium on mining in the Protected Zone is lifted in 1995, royalties from any mineral development would be subject to negotiation. There is at present no provision for a royalty or tax to be paid to the region or to local people (Arthur 1990).

2.5 Implications of the *Mabo* decision

The High Court's decision in the *Mabo* case has overturned the legal doctrine of *terra nullius*, rewriting the Australian common law to recognise indigenous title to land. The case dealt with the Murray Islands — three islands totalling less than 10 square kilometres in the eastern part of the Torres Strait. The decision brought to a close a ten year legal battle with the declaration that

> the Meriam people are entitled as against the whole world to possession, occupation, use and enjoyment of the lands of the Murray Islands.

Although the full implications of the *Mabo* decision cannot be known yet, it is a definitive turning point in Australian governance and indigenous policy.

The current Australian federal government seems already to have adopted a willingness to negotiate solutions. In such a context it should be possible to work out sensible accommodations for indigenous rights, and the ownership and management powers which will give them effect. In such a context MaSTS has much to offer, being a comprehensive proposal for a unique region and distinct culture. The Torres Strait presents an opportunity to rewrite white-indigenous relations in this country, on the basis of the accommodations and respect which did not exist in the 19th century, and on the basis of the new environment and sustainable development awareness of the late 20th century.

Issues to be addressed

3

MaSTS will address several issues of concern to Torres Strait Islander communities. Although it is unlikely that comprehensive solutions will be advanced for all of these complex issues, the Strategy will reflect Islander concerns and indicate where further investigations or policy changes are required.

MaSTS will:

- Establish locally-based institutional arrangements to develop and coordinate marine resource management and monitoring within the Torres Strait.

- Provide information relating to the natural environments of Torres Strait and advise on rational regional responses to global environmental change.

- Develop environmental planning and management guidelines, including provision of environmental impact appraisals to minimise the environmental costs of development.

- Provide practical solutions to current problems associated with locally generated pollution, including problems of inorganic or toxic household and industrial waste.

- Seek to reduce existing and potential threats of marine pollution or contamination at the regional scale (eg PNG mining operations and international shipping).

- Provide a forum through which the policies of the Torres Strait Islanders will be effectively represented in the management and utilisation of renewable resources.

- Facilitate the implementation of indigenous Islander rights in relation to Torres Strait Treaty re-negotiation, fisheries agreements, 'native title', and possible management tools such as 'sea closures', in light of the *Mabo* decision.

- Reconcile technological and traditional Islander management concepts and customs in a culturally acceptable marine strategy.

- Coordinate and further develop research programs jointly by outside researchers and local people to ensure cultural appropriateness, local content and relevance.

- Further develop and encourage education and training programs to enable Torres Strait residents to play a greater role in marine management and marine-related employment.

Some of these issues were discussed in detail in the Marine Workshop of the Surviving Columbus Conference in Darwin in October 1992 (see Jull *et al* 1993). Further work by the MaSTS team is continuing. ICC leaders, meanwhile, have been talking to federal and Queensland governments about their aspirations and needs in a new Torres Strait 'constitution' and administration. Some of the issues are at more advanced stages of investigation. These include environmental change and traditional conservation.

3.1 Natural environment

Massive environmental changes have occurred since the last phase of general climatic warming which began about 15,000 years ago. For Torres Strait these changes were particularly dramatic. For most of its history the region has been a bridge connecting the two landmasses of New Guinea and Australia. It is only since sometime between 8,500 and 6,500 years ago, when the postglacial rise of sea level flooded the Sahul shelf, that the Torres Strait became a marine environment, dotted with numerous small and isolated islands.

The Torres Strait has been described as

> the most ecologically complex area of one of the world's most extensive continental shelves ... Containing volcanic, continental, coral and alluvial islands, and fringing, platform and barrier reefs, the Strait offers a multitude of habitats and niches for the Indo-Pacific marine fauna, which itself has the greatest diversity of the ocean world (Nietschmann 1983, 134).

Four separate island groups are recognised:

An Eastern Group of high islands representing the eroded remnants of volcanic cones (eg Darnley and Stephens) and including the three islands of the Murray Group;

A Central Group of low sandy islands (cays) predominantly composed of reef-derived carbonate sand and surrounded by coral reefs;

A Western Group of high acid volcanic and granitic rocks, including Yam, Mabuiag, Moa, Badu, Thursday, Hammond, Horn, Friday, and Prince of Wales Islands;

A North-western Group of low islands where mangrove muds and alluvial sediments overlie reef-limestone and clay surfaces as at Saibai and Boigu Islands.

The Torres Strait is influenced by the tidal regimes of the semi-diurnal Coral Sea to the east and the diurnal regime of the Arafura to the west. In the Strait, tides are predominantly diurnal, with a maximum spring range of 3.5–4.0 m. Tidal current velocities often exceed 3m/sec within the shallow submarine topography around reefs and islands. The pattern is complex and seasonally variable and provides the dominant control over gross reef morphology as well as nearshore and onshore sedimentation patterns.

The region lies within a sub-humid savanna environment, with a dry season of 5.0–7.5 months duration and 95% of annual rainfall falling between December and April (Harris 1980). Seasonality is controlled by the north-south movements of the Intertropical Convergence Zone (ITCZ), and the vertical and longitudinal Walker and Hadley circulations. South-easterly trade winds dominate during the dry season, from May to October, and north-westerly moonsoon winds during the November to April wet season. Intermittant squally periods are associated with the north-westerlies winds which may exceed 50 km/hr for periods of up to 30 minutes. The highest magnitude storm wave events occur during the north-westerly wet season, and are most destructive or erosive when they occur during high tides. During the seasonal changeover phase, periods of negligible wind, known as the doldrums, occur.

Beaches, and the beachridge or dune systems landward of them, respond to seasonal wave movements by adjusting the direction of sediment movement. In unmodified beach compartments sediment is moved seasonally along the beach in two reverse directions, directly reflecting the wet and dry seasonal wind and wave directions. Sediments on the beaches and on the beachridges and foredunes move to the south and east during the monsoon season, with reverse drift occurring during the south-easterly season.

In coastal compartments which have been modified by clearing mangroves and other sediment-trapping vegetation, constructing buildings or dumping solid refuse on the mobile back-beach sediments, or by building stable ramps, seawalls, groynes or breakwaters within the tidal zone, this natural pattern of sand movement is interrupted. The most common result of this is that sediment movement occurs in one direction, with no reverse drift to replenish that sediment and an overall reduction in available sediment occurs within the beach compartment.

Kidd (1985, 106) described the usual response to this very common situation.

> Unfortunately a common response to such events is the construction of barriers immediately in front of the foredune. Typically these are solid, near vertical seawalls. Not only do these structures prevent the storm waves utilizing the emergency sandstore, their composition and inclination commonly results in almost all the incident wave energy being reflected back from them. This contrasts dramatically with a natural beach surface, which commonly absorbs much of the incident energy. In addition the crests of the reflected waves often coincide with those of the incoming waves, turbulence is greatly increased and the resultant erosion is much more severe as waves expend their energy by cutting down further into the beach. In extreme cases all the beach sand may be removed exposing bare rock surfaces to the waves. Seawalls may even be undermined by wave attack at their base and water from waves overtopping them may also percolate downwards removing the sand behind them. The walls collapse and the waves proceed to attack the now unprotected foredune.

Examples of the 'extreme case' can be seen within Torres Strait on the islands of Boigu and Saibai. Through MaSTS, practical advice and guidelines will be developed to ensure more appropriate responses are taken in relation to coastal protection.

Tropical cyclones are infrequent in the Torres Strait: only six cyclones having passed through the region during the last fifty years (Laurensz 1981). Nevertheless, cyclones and associated low atmospheric pressure storm surges in the ocean areas to the west and east can cause intense rainfall and extremely stormy wave conditions in the shallow waters of the Strait.

Wet season rainfall averages 1,600mm at Thursday Island, but the monthly total for the wettest month, January (mean 430mm), ranges between 167–863 mm (Barham 1985). Relatively drier areas occur on the leeward side of the larger high islands, but prolonged droughts are rare. They generally coincide with the El Nino-Southern Oscillation (ENSO) phenomenon — an event which occurs approximately every 5 to 7 years. Storage of fresh water supplies (as groundwater lenses or in constructed tanks) to last through the long dry season is, however, a major problem on many islands.

Temperatures in the region are not extreme and show relatively little day to day and seasonal veriation owing to the

moderating effect of the ocean. Mean temperatures range from about 26°C to 31°C, and it is rare to experience temperatures above 35°C.

3.2 Coastal development activities

In 1988, the Report of the Interdepartmental Committee (IDC) on the Torres Strait Islands acknowledged that

> generally, the standard of infrastructure in the Torres Strait is well below that normally accepted in mainland communities (O'Rourke 1988, 6.2).

As an immediate solution to some of the identified problems, the Minister for Aboriginal Affairs released the 'Priority Community Development Strategy for Torres Strait' (PCDS) with $23m to be spent over a three year period. Key elements of community infrastructure, notably landing craft unloading facilities, wharfs and sewerage systems, were immediately built and/or upgraded after a minimum of investigative research into their potential environmental impacts.

On the tiny islands of the Torres Strait, such operations, and their siting and ancillary requirements (eg excavation, earth moving, blasting, dredging, land reclamation, and disposal of wastes) have produced a variety of environmental problems, some of immediate effect and some with longer term impacts on subsistence livelihoods (see Mulrennan 1992).

The environmental costs of recent developments are considerable. For example, on Warraber Island — an elliptical coral cay, only 2–3 m above mean sea level and less than 750 m wide — the airstrip bisects the island along a north-west / south-east trajectory, providing a wind and wave corridor under storm surge conditions associated with the north-west monsoon. The provision of an airstrip on Yam Island has been responsible for a substantial breach in the coastal mangrove belt — one of the most biologically productive of all ecosystems and an important buffer against flooding and saltwater intrusion. While at Kubin, on Moa Island, and on other islands, extensive improvements to community wharf facilities and barge ramps have resulted in increased sedimentation and degradation of adjacent shallow reef habitats.

Insufficient advance thought appears to have been given to these adverse environmental impacts, with the result that they now threaten to increase the effects of coastal change and sea level rise (see Section 3.4). The Islanders are all too aware of these problems, acknowledging that

> Insensitive construction, even of essential infrastructure, can lead to unwanted but avoidable ill effects (Getano Lui, UNCED Earth Summit 1992, 3).

Development need not and should not occur to the detriment of the natural environment — which is why the ICC has firmly endorsed the principles of ecologically sustainable development as part of MaSTS. Basic ecological guidelines, to be incorporated within the Regional Development Planning process through MaSTS, will assist in minimising the environmental costs of future development.

The Horn Island gold mine provides another example of the inadequacy of existing mechanisms regarding the environmental impacts of coastal developments. As mentioned above (Section 2.1.3), the Commonwealth was not consulted over the decision to grant the mining lease nor did it review the final conditions for environmental protection. Furthermore despite repeated requests from the ICC in 1987, the Environmental Impact Assessment Report was not available for public comment.

The company concerned, Torres Strait Gold Pty Ltd, ceased mining on Horn Island in late 1989 due to bankruptcy, leaving a considerable area of disturbed ground including large pits and waste dumps. Rehabilitation of the mine site is currently in its early stages — but the possible long-term effects on the natural ecosystem from contamination by seepage and leachates from mine waste and from removal of sand from Horn Island beaches for use in construction works is a costly legacy to a development that also failed on economic grounds. The value of the mine to the local economy has been estimated to have been only about $2–3 m per year (Babbage 1990).

3.3 Marine pollution

Marine pollution is a major concern at a regional level in Torres Strait. Although the Torres Strait Baseline Study is dealing specifically with potential heavy metal pollution from the Fly River in PNG, there are many other sources of marine pollutants.

The concerns of the Islanders were recently voiced by former ICC Chairman, George Mye, who stated:

> From our islands we see lots of stuff floating by from the Fly River; it's what we don't see that worries us! (Mye, quoted in Staples 1992).

The present ICC Chairman, Getano Lui, in his submission to the UNCED Earth Summit in June 1992 mentioned several other developments in Indonesia and PNG which are of concern to the Islanders.

> Oil exploration in the Gulf of Papua and oil loading facilities at Kutubu on the Gulf of Papua, vessel traffic in the Strait, a pulp mill at Merauke in Indonesia, lead-zinc loading in the Gulf of Carpentaria, and oil development in the Timor Gap all cause worry in our Islands (Getano Lui, UNCED Earth Summit 1992, 3).

As yet, PNG has not developed effective oil spill prevention and clean up capacity, although funding arrangements for the extension of the OSSM oil spill prediction model have been finalised between AMSA and the PNG government, with the assistance of the Australian International Development Assistance Bureau (AIDAB).

For Australia's part, oil spill contingency plans have been established at national, state and regional levels. GBRMPA has developed REEFPLAN to protect the extremely vulnerable Great Barrier Reef and preliminary steps have been taken to extend this plan further north to cover the Torres Strait region, which is biophysically very similar.

A Marine Pollution Act has recently been proposed to replace the existing *Pollution of Waters by Oil Act 1973* and afford more complete protection to Queensland marine environment, in particular from pollution originating from all classes of ships and small craft. The Marine Pollution Act will incorporate the *International Convention for the Prevention of Pollution from Ships 1973/78* (MARPOL), which has been ratified by the Australian government.

Concerns about the potential risks of a major oil spill from international shipping make headlines; however, it is estimated that chronic, operational, small spills from shipping and handling of oil account for more than 70% of oil in the marine environment, while dramatic, large accidental spills contribute less than 10% (Carpenter & Maragos 1989). Thus, safety measures related to local fuel consumption and energy developments must be developed alongside regional oil spill contingency plans.

Increasing risks of pollution are also associated with increased sedimentation in the region due to land use changes, mining and construction, and the disposal of domestic and non-domestic wastes (including sewage and solid waste disposal). Such issues will be integrated within MaSTS.

Furthermore, the perception that the marine environment is being badly polluted will have as profound social and cultural effect as its physical manifestations are ever likely to. As a result it is imperative that Islanders are kept informed in a meaningful way of the results of all investigations into marine pollution. MaSTS will assist in the dissemination of such information.

3.4 Environmental change and projected sea level rise

The issues of global warming and associated sea level rise are of direct concern to Torres Strait Islanders. The best worldwide scientific predictions of global climate change, and the likely impacts of such changes, were collated by the Intergovernmental Panel on Climate Change (IPCC), and have since been published and distributed widely (Houghton *et al* 1990). For tropical maritime environments, such as Torres Strait, an average temperature rise of 1.5 to 2 °C is likely by the year 2100, with sea level predicted to rise by 65±30 cms during that time. Weather patterns are likely to become more unstable, and extreme weather events more common in association with those changes.

No specific studies have been carried out on the Torres Strait Islands to determine the likely impacts of projected global climate change and sea level rise. However, a sufficient number of detailed studies have been carried out for the northern Australian mainland coast, the PNG coastline and for similar islands in the south-west Pacific to identify in general terms the likely impacts on each group of Torres Strait Islands.

Both the Eastern and Western Groups (see Section 3.1) will be affected by sea level rise in ways similar to those described for Misima and Lihir Islands in PNG (Bualia & Sullivan 1990) or Viti Levu and Beqa Island in Fiji (Nunn 1990). The areas most vulnerable to sea level rise are backbeach dunes, older beachridges now forming coastal plains and small estuaries at the mouths of creeks or rivers. In economic terms these are common locations for village housing, for roads or tracks, bridges, boat ramps and other infrastructure, which is generally constructed on flat coastal land. The proportional loss of land from such islands is likely to be small, but its value may be relatively high. Conservative planning would suggest that future construction should be located well back (more than 50m) from the shoreline, and that active planting and maintenance of vegetation which stabilises the coastal sediments should be encouraged.

The North-western Group of low-lying, muddy, deltaic islands includes the islands of Boigu and Saibai. Between them is the small island of Dauan — a granitic high island — which will be affected like those of the Western Group (above). Boigu and Saibai, like Daru Island in PNG, are deltaic floodplain landforms, and will respond in ways described by Bualia (1990) for the Kikori area, Hughes and Bualia (1990) for the Sepik plains, and

Bualia and Sullivan (1990) for deltaic floodplains and lagoon and sandspit coastal areas more generally.

Apart from extended seasonal flooding and eventual loss of the lowest lying land, such areas may undergo major changes in the vegetation they support. In addition, a relatively early impact will be the salinisation of fresh groundwater supplies and the subsequent abandonment of wells and bores as the quality of drinking water and water for gardens deteriorates. For coastal land and islands where subsistence gardening is still practised, this may result in a direct economic loss. In other areas there will be a need to change to more expensive water supplies such as roof runoff tanks, and consequently sheet-metal roofing for houses.

The Central Group is likely to be affected in the ways described for atoll and other coral-based islands such as the Marshall Islands, Tuvalu, Kiribati, the Maldives, the Trobriand group in PNG and several hundred small tropical islands around the world (Sullivan & Pernetta 1990; Sullivan & Gibson 1991).

The impacts of even a small rise in sea level on these low-lying islands will be severe. The major impacts include contamination of the fresh groundwater lens which floats above the seawater on such islands, and loss of coastal land under the influence of higher waves. Lack of freshwater will make many low-lying islands uninhabitable before there is a severe loss of land. Provision of alternative sources of drinking and gardening water, and careful maintenance or establishment of coastal fringe vegetation will delay the impacts of such sea level rise.

In common with other tropical areas, on all of the Torres Strait Islands there are strong implications for human health resulting from climate change (McGregor 1990; Nurse 1990). These include problems of heat stress and thermal comfort, especially for people working outdoors, increases in parasitic diseases including malaria, filariasis (elephantiasis), various forms of encephalitis, and many gastric disorders. In areas where the delivery of health services is already poor, increased health risks due to climate changes become potentially more serious.

Through MaSTS, more detailed work will be carried out to identify vulnerable or threatened sectors of the coastal zone of the main inhabited Islands. Specific mapping to indicate areas most likely to suffer the impacts of increased wave heights or saline groundwater intrusion will enable better management planning for coastal infrastructure or other proposed developments, and will enable the identification of more stable alternative areas for infrastructure.

In addition to global warming and sea level rise, global problems related to the continued effects of rising human populations, ozone depletion and atmospheric change should also be built into any strategic planning process. Predictions for the future are always uncertain and difficult to apply in the short term; nevertheless, lack of scientific certitude should not be used as a reason for postponing measures to deal with environmental degradation.

3.5 Customary marine tenure and traditional resource use

The concept of customary marine tenure is a particularly valuable marine conservation measure. It is also essential to the cultural security in the short term, and survival in the long term of the Torres Strait Islander culture. Islanders have always had their own customary tenure laws. (Customary land tenure was recognised in principle in the *Mabo* decision.) Research on customary tenure and its applications is underway, with the MaSTS team working with specialists on the mainland and from overseas to obtain precedents for the development of Torres Strait options. An equitable and workable system must be devised for present-day circumstances. This matter will be a critical subject for political negotiation between Torres Strait Islanders and the Australian and Queensland governments.

3.5.1 Conservation as resource management

Traditional societies maintain a strong conservation ethic, but it is philosophically very different from Western concepts of conservation. In traditional societies the conservation concept is allied to the present wider Australian philosophy of sustainable resource use. Such a concept may be somewhat different from that of ecologically sustainable development, which has embedded within it the view that natural ecosystems should be given consideration simply because they exist, or because of biodiversity values. In a western conservationist model, values are placed on environmental attributes. These values are based on such factors as rarity or uniqueness, representativeness, threat, and diversity (see for example Margules & Usher 1981). In traditional societies values relate more closely to the use which is made of any particular resource, and resource management strategies are based on how to both use and maintain the supply of highly valued resources.

Established traditional practices of marine resource use are well developed in the Pacific atoll states where land resources are minimal. There are strong similarities in resource availability and traditional resource use between the Central Group of the Torres Strait Islands and Pacific atoll states such as Kiribati and Tuvalu (formerly the Gilbert and Ellice Islands). The development and implementation of MaSTS will draw on experience from these and other settings.

3.5.2 Traditional management practices and Torres Strait

Like the world's other indigenous peoples, Torres Strait Islanders have had their own well-established views and systems of resource management. Traditional resource management in the past occurred in relative isolation at a time when pressures were generally not so great on the environment. Recent challenges to the marine environment — including shipping, mining and commercial fisheries — are not being met by traditional management practices and there is an urgent need to assess how best to integrate modern technical knowledge within traditional systems of management.

Of particular relevance to this integration process will be the material collated by Nietschmann (1985), Cordell (1991) and Johannes and MacFarlane (1991) on customary marine tenure (CMT) in Torres Strait, and its impact on marine conservation. While the information presented is generalised, and individual instances which do not conform to the general pattern can be identified, these data are an excellent basis for further elucidating the persistence of CMT practices, and their significance in contributing towards a management strategy. Discussions with Torres Strait Islanders to ascertain their views on CMT and its possible application to MaSTS will provide a major component of this work.

3.5.3 Depletion of marine resources

Conservation and management of marine resources is essential to the Islanders. Commercial fisheries provide one of the few sectors of the Australian economy open for their participation. The link with the sea is central to the maintenance of certain Islander customs and traditional practices.

At present the motivation to exploit is greater than the power of local control over marine resources. The Islanders see effective management of marine resources as dependent on selective exploitation of local resources by a small community of people, combined with local control over home reefs and over sea space in the immediate vicinity of the inhabited islands.

3.5.4 Marine management strategies and economic independence

Access to income is fundamental to self-reliance. What is the acceptable limit of commercial exploitation of the Torres Strait marine environment, who owns the resource, and who should benefit from this? These are issues which the next stage in the development of MaSTS will consider (see Section 3.6).

Underpining the conservation management aspects of MaSTS is the assumption that marine resource management must contribute to the long-term physical and social sustainable development of the Torres Strait Islanders. The Strategy must be directed towards ecologically sustainable resource use, not simply species preservation or other forms of conservation. Only Torres Strait Islanders necessarily have an interest in long-term sustainable resource use. Any long-term strategy for managing these resources must be designed and implemented by the Islanders, and they must have the legal authority to be able to do so.

3.5.5 Traditional management and education

It has been noted throughout the Pacific that as access to the cash economy increases, traditional practices and customary regulations diminish. Traditional practices are no longer always passed on between generations, and memories become less clear. One aspect of the MaSTS program should be to ensure that traditional knowledge of marine tenure and practices is incorporated into school curricula, as recommended by Johannes and MacFarlane (1991). In primary school courses dealing with the natural environment there should be a concerted attempt to include curriculum materials which deal with Torres Strait traditions, including the traditions associated with the exploitation of marine resources. In this way social, cultural and economic information can be integrated, and the rich culture of the Islands maintained.

Both the traditional knowledge of the Islanders and that which has been recorded as a result of studies by Nietschmann (1985) and Johannes and MacFarlane (1991) can be used to develop an appropriate marine management strategy. Traditional concepts

such as those of changing territory boundaries in response to changes in tides and winds, access rights, and knowledge of habitats and marine animal behaviour must all be incorporated into MaSTS.

3.6 Fisheries

Fisheries resources are a major source of contention and concern in Torres Strait. Traditional fisheries, community fisheries, licensed commercial fisheries and recreational fisheries must all be considered in an overall environmental management plan. Agreement on any changes to existing legislation or regulated practices will involve the Australian government, Queensland State government and the government of the independent state of Papua New Guinea.

The CSIRO Fisheries Division and the Queensland Department of Primary Industries have completed several detailed investigations directed towards the management of fisheries resources in Torres Strait, and have several resource inventories and management plans underway. University researchers, particularly from James Cook University, and scientists from GBRMPA and the Australian Institute of Marine Science (AIMS) have also been involved in aspects of fisheries research in the Strait and surrounding marine areas.

Elmer and Coles (1991) have summarised the existing consultative and catch sharing arrangements for Torres Strait, in addition to highlighting the complexity of economic and social issues which concern fisheries managers.

These many aspects of marine resource inventory, and economic and fisheries resource management will form a major component of the ongoing MaSTS process. Rather than trivialise these complex matters here, fisheries management issues will form a major part of the next stage in the development of MaSTS. A workshop, to be held on Thursday Island in June 1993, will seek appropriate management advice on strategies for immediate implemenation. Workshop participants will include invited representatives of organisations already involved in fisheries research and management in Torres Strait.

Implementation strategies

4

The Marine Strategy (MaSTS) will be implemented through a series of **action plans**. These will be developed from the strategies outlined below. These strategies seek to establish the broad framework for implementation.

4.1 Research and monitoring

Successful resource management policies and strategies depend on adequate scientific understanding of the nature of those resources. Research and monitoring will, therefore, provide an essential component of the development and implementation of MaSTS.

Research in the Torres Strait has generally concentrated on studies of specific species, isolated components, or limited areas within the Strait. Despite the value of much of this work, knowledge gained from a broader scientific program is required for application in formal planning and decision making. Research efforts will, therefore, require coordination and integration to ensure that interdisciplinary and interagency studies maximise the utility of the results and minimise the costs of producing the advice.

To a large extent, the success or failure of renewable resource management and marine conservation in the Strait hinges on the quality of the biological advice. Prolonged harvests above the limits defined by the biological system will inevitably lead to collapse of the resource. Thus biological assessment must be the first consideration on which overall management advice is based.

Increasing pressure on northern Australia as a source of non-renewable resources creates an urgent need to understand its tropical ecosystems and to minimise the negative impacts of resource development. At the same time the productivity of the renewable resource base upon which Islanders depend must be maintained and/or enhanced. Above all, development planning must recognise and incorporate the viewpoint and knowledge of the Islanders themselves.

- *The ICC will establish a Torres Strait Advisory Committee which will:*
  - *provide scientific advice, consult with relevant experts and direct the research efforts required for the successful development and implementation of MaSTS;*
  - *integrate traditional knowledge as a fundamental element of the ongoing scientific process.*

4.2 Shared management

Successful management requires sharing. The nature of political jurisdictions and ocean systems themselves require this. Sharing in management — cooperation — is also likely to produce the best results for all participants. Stakeholders, with responsibility for decision making about resource utilisation, include Islanders and other residents, governments and commercial fishermen. Government fisheries authorities in the region already have liaison/decision making procedures in place, which include Islander representation. MaSTS will build upon and seek to maximise Islander involvement in existing and emergent structures. Conservation objectives in Torres Strait can only be achieved if the people who depend on the living resources have an active role in their management.

Shared management will be guided by the principles of MaSTS (Section 1.4). The subjects of cooperation will include sustainable use of harvestable stocks, protection of endangered stocks, restoration of depleted stocks, protection and improvement of habitat, reduction of waste, and identification and evaluation of opportunities for economic development based upon renewable resources.

- *The ICC will encourage coordination with government departments in the establishment of shared management processes, involving all resource users in decisions about renewable and non-renewable marine resource management.*

4.3 Integrated resource planning and management

4.3.1 Planning and management

Administration and management of natural resources are presently conducted on a single sector basis. This approach fails to acknowledge that activities in one sector (eg tourism and infrastructure development) have an impact on resources in other sectors (eg fisheries). The development of MaSTS will improve integration and coordination between resource sectors and seek to reconcile differences in objectives and to optimise use of resources, with due consideration given to all resource values — economic, social, cultural and environmental.

A framework is required to integrate the management of the Strait's marine resources. Such a framework should provide for shared participation by users, resolution of resource conflicts and harmonising activities flowing from different departmental and user group objectives. To this end, a land use plan (marine use plan) could be used to: anticipate, resolve and minimise competing demands or land use conflicts; optimise resource allocation; and, ensure integrated management of resources. The process would involve extensive public participation in order to ensure consideration of local needs.

- *The ICC will seek opportunities to resolve multiple resource conflicts in the Torres Strait marine area by participating in resource planning and management with local, state and federal agencies and resource users.*

- *The ICC will identify and recommend modification to existing institutional and legislative procedures which would ensure effective integration of marine management.*

- *The ICC will explore, with marine resource users, the benefits of a land use planning process as an integrating tool in marine areas.*

4.3.2 Sustainable development of renewable resources

Management of renewable resources in the marine environment should be effectively integrated to ensure the development and sustainability of social, cultural and economic benefits for Islanders. To do this, management should be designed to maintain, restore and enhance the renewable resource base. Management should also attempt to enhance and diversify the use of renewable resources, and to optimise the benefits from renewable resource use.

The development of management plans should be based on adequate data, where possible, and should be consistent with those existing species and habitat management plans considered to be effective.

- *The ICC, with the assistance of an Advisory Committee and consultations with resource users, will:*
  - *establish priorities for the development of detailed management plans (eg species specific management plans for marine mammals and fish);*
  - *ensure that socio-economic development strategies for the Torres Strait are consistent with MaSTS.*

4.3.3 Non-renewable resource development

Non-renewable resource developments, including marine transport, must be planned with full consideration of possible effects on other resources and resource users. In situations where non-renewable resource developments are already under way, the effects of these developments will be monitored. Those developments that are compatible with marine conservation should be encouraged.

- *Industry must include measures to protect the Torres Strait marine environment and resources when planning its activities.*
- *The Torres Strait Baseline Study will provide an assessment of existing environmental effects of PNG mining operations on the Torres Strait marine environment and provide guidelines for on-going monitoring of the marine environment.*
- *Further measures will be taken to encourage the development of adequate marine transportation regulations and oil spill contingency plans.*

4.3.4 Environmental impact assessment

Terms and conditions, including the possibility of new legislation, must be established to minimise the adverse effects of resource development on the Torres Strait marine environment. This approach is most effective if potential environmental impacts are identified early in the project design phase.

- *Environmental assessment procedures will be reviewed and refined.*
- *Environmental criteria and guidelines will be developed to aid in determining the significance and acceptability of particular environmental impacts.*

- *Improved project monitoring and research into mitigative strategies will be encouraged.*

4.4 Marine environmental quality

The quality of the Torres Strait marine environment must be maintained, protected and enhanced to ensure the well-being of its marine life and human population. Conservation of marine renewable resources is best accomplished within an integrated resource management system. It may be necessary to create a network of protected areas as a component of this system. Such a network should include areas designated for cultural, scientific, educational, aesthetic, recreational or biological purposes, or to maintain natural conditions or preserve certain features. Long-term management plans for each marine protected area should be formulated in the context of broader regional and species management plans and be consistent with and supportive of overall renewable resource management objectives.

- *Maintenance, protection and enhancement of the quality of the Torres Strait marine environment will depend upon:*

  – *development of marine environmental quality criteria for the Strait's marine area;*

  – *promotion of national and international initiatives to reduce long-range transport of contaminants;*

  – *establishment of a comprehensive program for pollution control, including pollution from land-based sources should be developed; and*

  – *selection of marine areas requiring special protection.*

4.5 Public knowledge

Awareness of the benefits of marine conservation enables policy makers, governments, developers and the general public to achieve conservation objectives.

Information should be widely disseminated through informal education programs directed at adults and formal education programs for school children. Mass media (radio, newspapers, television) should also be fully used to reach as many people as possible. It would be useful and feasible to develop a computer atlas.

Islander traditional knowledge of the marine environment must be preserved and utilised in the conservation and management of the environment.

Effective management is dependent on the involvement of resource users at the community level. Additional benefits are gained if management personnel come from local communities. Thus the provision of relevant education and training programs (for example, a ranger program) is needed.

Many of the concepts and management strategies which will be developed are derived from fields of fisheries biology, oceanography, geomorphology, ecology and anthropology and are complex and theoretical. Relevant scientific concepts should be formulated in practical terms that are appropriate to local managers and local resource users.

- *The ICC will develop processes to promote opportunities for communication of MaSTS activities to Torres Strait Islanders, to governments, public bodies, industry, academic and research communties, and to the interested public.*

4.6 International cooperation

International cooperation is important for the conservation of living resources and the protection of the environment (Section 2.2).

Australia is a party to several international agreements which involve commitments to conserve marine renewable resources and ecosystems. These include the World Conservation Strategy, Agenda 21 of the UNCED Earth Summit, the Convention on International Trade of Endangered Species, etc. When ratified by Australia, the United Nations Convention on the Law of the Sea (UNCLOS) will bring additional obligations.

Torres Strait falls within the area of the South Pacific Regional Environment Programme (commonly known as the SPREP). Australia and PNG are partners, along with their South Pacific neighbours, in the development and implementation of the Programme and many of the articles of the Convention are relevant to the environmental issues of the Torres Strait.

Bilateral management of shared stocks of fish and mammals is essential to ensuring the sustainable utilisation of the Strait's marine renewable resources. Similarly, cooperative research initiatives can make a significant contribution to Torres Strait marine conservation.

Another important aspect of international cooperation is the progress in other parts of the world in indigenous development, land/sea claims settlements, self-management and self-government, and marine and related environmental strategies. For example, indigenous marine arrangements and projects in Canada, Alaska and northern Norway have many relevant lessons (eg Mulrennan & Jull 1992). In each of those regions governments and indigenous peoples have developed new relationships with each other thanks to indigenous initiatives on coastal and marine issues.

- *The ICC will ensure that the objectives of MaSTS consider Australia's commitments.*

- *Cooperative management of shared stocks of fish and mammals will be encouraged between PNG, Indonesia and Australia.*

- *Comparative studies and exchanges with other 'first world' countries will be encouraged where relevant to:*

  - *the resolution of indigenous coastal and marine issues, or environment and resource management issues more widely; and*

  - *cooperation of indigenous peoples and senior governments in regional development.*

4.7 Monitoring and review

MaSTS will be implemented in such a way that its activities are consistent with those of related environmental management strategies.

MaSTS and subsidiary activities undertaken will be evaluated periodically and revised whenever necessary to ensure that responsible groups are undertaking their actions properly and that a record is kept of the effect of those actions on stocks of resources, on the environment and on people.

MaSTS and its associated action plans will evolve over time. Thus, MaSTS will be modified when new knowledge becomes available through monitoring or evaluation processes, or as a result of other developments.

- *ICC, with the assistance of the Advisory Committee, will ensure that the implementation of MaSTS is periodically monitored, evaluated and revised whenever necessary.*

4.8 Immediate action plan

1) This document to be widely circulated and public comment invited.

2) Advisory Committee to MaSTS to be established.

3) Discussions with DFAT concerning the Torres Strait Treaty and its role in the protection of the Torres Strait marine environment.

4) Workshop on 'Marine management opportunities and constraints: options for change' to be held on Thursday Island in mid June 1993. Invited speakers will present management options for particular resources, provide environmental planning guidelines, outline immediate research needs, and advise on *Mabo* related action in relation to customary marine tenure.

5) Publication of workshop proceedings as the basis for a series of immediate management action plans.

6) In consultation with relevant Queensland government departments, a handbook of advice on environmental assessment for planners and developers to be produced for wide distribution.

7) Pending funding being available from the Australian government, a Torres Strait environmental resource centre to be established in association with the ICC office on Thursday Island. The centre will be staffed by the MaSTS director and adminstrative support staff. Trainees in CEPANCRM and ATSIC training schemes will be supervised through the centre. The primary function of the centre will be to coordinate and facilitate environmental planning and management activities in the Torres Strait. Liaison between ICC, the MaSTS team, the Advisory Committee and all government departments to be through this coordinating structure.

Transition to MaSTS

5

The transition from a hodgepodge of policies and programs to a regionally centred comprehensive approach will not happen overnight. It requires significant adjustments to the thinking of officials, users, scientists, Islanders, and others. The solutions may not be made easier by the fact that Australian governments themselves must revise their policies in light of the *Mabo* decision and in light of nationwide and international moves to greater indigenous self-government. On the other hand, *Mabo* has provided a reason for change, one for which governments might have otherwise lacked political will.

The environment and the economic base of Torres Strait, as addressed by this Marine Strategy for Torres Strait (MaSTS), is of no less importance for the long-term future of Islanders than regional government and service delivery — indeed, they are intertwined. governments often believe that social programs alone will 'solve' the problems of indigenous peoples, but the experience everywhere in the 'first world' has shown they are wrong (Jull 1991). Cultural autonomy, significant measures of self-government, and rights to ownership and management of traditionally used lands, waters and resources form an indissoluble package of necessary measures.

Australia has signed various international documents and made numerous statements through federal prime ministers and ministers committing itself to a new direction in policy towards Torres Strait Islanders and Aborigines. Australia has also signed environmental and sustainable development commitments such as the documents of the Rio Earth Summit of 1992. The Torres Strait is a relatively small region and population. It can serve Australia with little difficulty as a model for new policy thinking and the removal of indigenous hardships and grievances. This year, 1993, being the International Year of Indigenous Peoples, would be a good time for commitment by government to such change.

References and further reading

The following bibliography contains a complete listing of all references cited throughout the document. Additional references are also included which the MaSTS team have found useful in putting the document together. The references have been grouped under various sub-headings as a general guide to the reader but these categories are by no means exclusive.

General reading

Beckett JR, 1987. *The Torres Strait Islanders: custom and colonialism*, Cambridge University Press, Cambridge.

Lawrence D & Cansfield-Smith T (eds), 1991. *Sustainable development for traditional inhabitants of the Torres Strait region*, GBRMPA Workshop Series 16, AGPS, Canberra.

Singe J, 1979. *The Torres Strait*, University of Queensland Press, St Lucia.

Environmental studies

Amin M, 1978. A statistical analysis of storm surges in Torres Strait, *Australian Journal of Marine & Freshwater Research* **29**, 479–96.

Barham AJ, 1981. Landuse and environmental change in the Western Torres Strait Islands, northern Queensland, Australia, Field-Work Report, Department of Geography, University College London and Institute of Archaeology, University of London, unpublished.

Barham AJ & Harris DR, 1983. Prchistory and palaeoecology of Torres Strait, *in* PM Masters & NC Flemming (eds), *Quaternary coastlines and marine archaeology: towards the prehistory of land bridges and continental shelves*, Academic Press, London, 529–57.

Barham AJ & Harris DR, 1985. Relict field systems in the Torres Strait region, *in* I Farrington (ed), *Prehistoric intensive agriculture in the tropics*, BAR International Series 232, British Archaeological Reports, Oxford, 247–83.

Bonnell M & Gilmour DA, 1980. Variations in short-term rainfall intensity in relation to synoptic climatological aspects of the humid tropical northeast Queensland coast, *Singapore Journal of Tropical Geography* **1**(2), 16–30.

Brundtland GH *et al*, 1987. *Our common future: the report of the World Commission on environment and development*, Oxford University Press, Oxford.

Bualia L, 1990. The impacts of sea level rise on a low-lying landscape in Papua New Guinea: a case study from the Gulf of Papua, *in* Pernetta & Hughes (eds), 224–37.

Bualia L & Sullivan ME, 1990. The impacts of possible global warming generated sea level rise on selected coastal landforms in Papua New Guinea, *in* Pernetta & Hughes (eds), 193–9.

Connell J & Lea J, 1992. My country will not be there: global warming, development and the planning response in small island states, *Cities* **9**(4), 295–309.

Earthscan (ed), 1991. *Caring for the Earth: a strategy for sustainable living*, IUCN, UNEP & WWF, Gland, Switzerland.

Easton AK, 1970. *The tides of the continent of Australia*, Research Paper No 37, Horace Lamb Centre for Oceanographical Research, Flinders University of South Australia, Bedford Park.

GBRMPA, 1992. A 25 year Strategic Plan for the Great Barrier Reef World Heritage Area 1992–2017, Draft document, Great Barrier Reef Marine Park Authority, Townsville.

Haddon AC, Sollas WJ & Cole GAJ, 1894. On the geology of Torres Straits, *Transactions of the Royal Irish Academy* **30**(11), 419–76.

Harris DR, 1977. Subsistence strategies across Torres Strait, *in* J Allen, J Golson & R Jones (eds), *Sunda and Sahul: prehistoric studies in South-east Asia, Melanesia and Australia*, Academic Press, London, 421–63.

Harris DR, 1979. Foragers and farmers in the western Torres Strait Islands: an historical analysis of economic, demographic and spatial differentiation, *in* PC Burham & RF Ellen (eds), *Social and ecological systems*, Academic Press, London, 75–109.

Harris DR (ed), 1980. *Human ecology in savanna environments*, Academic Press, London.

Henderson RA & Stephenson PJ (eds), 1980. *The geology and geophysics of northeastern Australia*, Geological Society of Australia, Queensland Division, Brisbane.

Hopley D & Harvey N, 1978. Regional variation in storm surge characteristics around the Australian coast: a preliminary investigation, *in* RL Heathcote & BG Thom (eds), *Natural hazards in Australia: proceedings of a symposium sponsored by Australian Academy of Science, Institute of Australian Geographers, Academy of the Social Sciences in Australia*, Australian Academy of Science, Canberra, 164–85.

Houghton JT, Jenkins GJ & Ephraums JJ, 1990. *Climate change: the IPCC scientific assessment*, Intergovernmental

Panel on Climate Change, Cambridge University Press, Cambridge.

Hughes PJ & Bualia L, 1990. Murik Lakes and the mouth of the Sepik River, *in* Pernetta & Hughes (eds), 243–6.

Hughes PJ & McGregor GR (eds), 1990. *Global warming-related effects on agriculture and human health and comfort in the South Pacific*, Association of South Pacific Environmental Institutions and UPNG, Port Moresby.

ICC & MaSTS Team 1992. Report on preparatory meeting of the ICC and MaSTS Team for the Marine Workshop of Surviving Columbus Conference, Beaufort Hotel, Darwin, September 1992, unpublished.

IPCC Coastal Zone Management Sub-Group, 1990. *Strategies for adaptation to sea level rise*, IPCC Response Strategies Working Group, Geneva.

Jongsma D, 1970. Eustatic sea level changes in the Arafura Sea, *Nature* **228**, 150–1.

Jull P, Mulrennan ME & Sullivan ME (eds), 1993. *Surviving Columbus: indigenous environments today and tomorrow*, Proceedings of conference organised by the NARU, ICC, NLC and CLC, October 1992, forthcoming.

Kidd RW, 1985. Coastal zone management: why are our beaches disappearing?, *Ples* **1**, 103–11.

Lourensz RS, 1981. *Tropical cyclones in the Australian region July 1909 to June 1980*, AGPS, Canberra.

Lui G, 1992. The Torres Strait Islands of Australia, paper prepared for the UNCED Earth Summit, Rio de Janeiro, June 1992.

Maclean RF, Stoddart DR, Hopley D & Polach H, 1978. Sea level change in the Holocene on the northern Great Barrier Reef, *Phil. Trans. R.Soc. Lond. A.* **291**, 167–86.

McGregor GR, 1990. Possible impacts of climate change on human thermal comfort in the western tropical Pacific, *in* Hughes & McGregor (eds), 98–113.

Moore DR, 1984. *The Torres Strait collections of AC Haddon*, British Museum, London.

Nunn P, 1990. Potential impacts of projected sea level rise on Pacific Island states (Cook Islands, Fiji, Kiribati, Tonga, and Western Samoa): a preliminary report, *in* Pernetta & Hughes (eds), 127–48.

Nurse GT, 1990. Possible effects of climate change on health in the Pacific Island countries, *in* Hughes & McGregor (eds), 114–21.

Pernetta JC & Hughes PJ (eds), *Implications of expected climate changes in the South Pacific region: an overview*, UNEP regional seas reports & studies No.128, Nairobi, Kenya.

Sabitay A, 1978. The status of hydrocarbon exploration, offshore northeastern Australia and the Gulf of Papua, *Austr. Petrol. Explor. Assoc. Jnl.* **18**, 77–92.

Staples J, 1992. Torres Strait: islands at risk, *Habitat Australia* **20**(2), 26–30.
Sullivan ME & Gibson L, 1991. *Environmental planning, climate change and potential sea level rise: report on a mission to Kiribati*, South Pacific Regional Environmental Programme Reports and Studies No.50.
Sullivan ME & Pernetta JC, 1990. The effect of sea level rise on atolls and motu, *in* Pernetta & Hughes (eds), 41–55.
Swan B, 1979. The presence of sand dunes in a tropical low energy zone, Friday Island, Torres Strait (Australia), *Revue de Geomorphologie dynamique* **28**, 61–72.
Van Andel TH & Veevers JJ, 1967. Morphology and sediments of the Timor Sea, *Bulletin [Australian] Bureau of Mineral Resources, Geology and Geophysics* **83**.
Van Andel TH, Heath GR, Moore TC & McGeary DFR, 1967. Late quaternary history, climate and oceanography of the Timor Sea, NW Australia, *Amer. J. Sci.* **265**, 737–58.
Vanderwal R, 1973. The Torres Strait: protohistory and beyond, *University of Queensland Anthropology Museum Occasional Papers in Anthropology*, No.2, 157–94.
Walker D (ed), 1972. *Bridge and barrier: the natural and cultural history of Torres Strait*, Department of Biogeography and Geomorphology, Australian National University, Canberra.
Willmott WF, 1972. *Daru-Maer, Papua and Queensland. 1:250,000 Geological Series explanatory notes*, AGPS, Canberra.
Wyrtki K, 1960. *The surface circulation in the Coral and Tasman Seas*, CSIRO Division of Fisheries and Oceanography technical papers No.8., Cronulla, NSW.

Customary marine tenure and traditional conservation

Allen BJ, 1987. Environmental ethics and village agriculture, *in* Stratigos S & Hughes PJ, *The ethics of development, Vol.1, The Pacific in the 21st century*, UPNG Press, Port Moresby, 7–20.
Bulmer RNH, 1982. Traditional conservation practices in Papua New Guinea, *in* Morauta *et al*, 59–77.
Carrier J, 1982. Conservation and conceptions of the environment: a Manus Province case study, *in* Morauta *et al*, 39–43.
Cordell J, [1991]. Managing sea country. tenure and sustainability of Aboriginal and Torres Strait Islander marine resources, Report on Indigenous Fishing, Ecologically Sustainable Development, Fisheries Working Group, Canberra, unpublished.

Cordell J & Fitzpatrick JM, 1987. Torres Strait: cultural identity and the sea, *CS Quarterly* **11**(2), 15–17.

Crocombe R, 1989. Tomorrow's customary tenure in relation to conservation, *in* PEJ Thomas (ed), *Report of the workshop on customary tenure, traditional resource management and nature conservation*, South Pacific Commission, Noumea, 21–8.

Davis S, 1985. Traditional management of the littoral zone among the Yolngu of north Australia, *in* Ruddle & Johannes (eds), 101–24.

Dwyer PD, 1982. Wildlife conservation and tradition in the highlands of Papua New Guinea, *in* Morauta *et al*, 173–89.

Haines AK, Williams GC & Coates D, 1986. *Torres Strait Fisheries Seminar, Port Moresby*, 11–14 February 1985, AGPS, Canberra.

Haiveta C, 1982. Freshwater supplies, past and present, in Lese Oalai, a Gulf Province village, *in* Morauta *et al*, 273–7.

Hudson BET, 1982. Dugong myth and management in Papua New Guinea, *in* Morauta *et al*, 311–15.

Johannes RE, 1978. Traditional marine conservation methods in Oceania and their demise, *Annual Review of Ecology and Systematics* **9**, 349–64.

Johannes RE & MacFarlane JW, 1991. *Traditional fishing in the Torres Strait Islands*, CSIRO Division of Fisheries, Hobart.

Margules CR & Usher MB, 1981. Criteria used in assessing wildlife conservation potential: a review, *Biological Conservation* **21**, 79–109.

Morauta L, Pernetta J & Heaney W (eds), 1982. *Traditional conservation in Papua New Guinea: implications for today*, Institute of Applied Social and Economic Research (IASER), Discussion Paper No.35, Boroko, PNG.

Nietschmann B, 1985. Torres Strait Islander sea resource management and sea rights, *in* Ruddle & Johannes (eds), 125–54.

Olewale E & Sedu D, 1982. Momoro (the dugong) in the Western Province, *in* Morauta *et al*, 251–5.

Paglau M, 1982. Conservation of soil, water and forest in the upper Simbu Valley, *in* Morauta *et al*, 115–19.

Ruddle K & Johannes RE (eds), 1985. *The traditional knowledge and management of coastal systems in Asia and the Pacific*, UNESCO, Jakarta.

Spring SC, 1982. Marine turtle conservation in Papua New Guinea, *in* Morauta *et al*, 303–6.

Tabai I, 1987. The ethics of development: a Kiribati view, *in* Stratigos S & Hughes PJ, *The ethics of development: the Pacific in the 21st century*, Vol.1, UPNG Press, Port Moresby, 40–49.

World Wildlife Fund for Nature (WWF), 1992. *Papua New Guinea Protected Areas Programme. Review of the management and status of protected areas, and action plan*, WWF and PNG Department of Environment & Conservation, Port Moresby.

Wright A, 1985. Marine resource use in Papua New Guinea: can traditional concepts and contemporary development be integrated?, *in* Ruddle & Johannes (eds), 79–99.

Zann LP, 1985. Traditional management and conservation of fisheries in Kiribati and Tuvalu atolls, *in* Ruddle & Johannes (eds), 53–77.

Regional development

Arthur WS (with assistance from V McGrath), 1990. *Torres Strait development study, 1989*, Australian Institute of Aboriginal Studies, Canberra.

Arthur WS, 1991. *Indigenous economic development in the Torres Strait: possibilities and limitations*, CAEPR Discussion Paper 4, ANU, Canberra.

Babbage R, 1990. *The strategic significance of Torres Strait*, Strategic and Defence Studies Centre, ANU, Canberra.

Elmer M & Coles R, 1991. Torres Strait fisheries management, *in* Lawrence & Cansfield-Smith (eds), 283–93.

Haines A, 1991. Policy context of the Torres Strait environment baseline study, *in* Lawrence & Cansfield-Smith (eds), 473–9.

Haines AK, Williams GC & Coates D, 1986. *Torres Strait Fisheries Seminar, Port Moresby*, 11–14 February 1985, AGPS, Canberra.

ICC 1991. Principles and objectives for the future of Torres Strait, attachment to submission letter to Premier Goss, May 16.

Jull P, 1991. *The politics of northern frontiers*, North Australia Research Unit, Darwin.

Kehoe-Forutan SJ, 1991. Self-management and the bureaucracy: the example of Thursday Island, *in* Lawrence & Cansfield-Smith (eds), 421–6.

Laffan J, 1991. The Torres Strait Treaty and the environment, *in* Lawrence & Cansfield-Smith (eds), 451–60.

Lawrence D & Cansfield-Smith T (eds), 1991. *Sustainable development for traditional inhabitants of the Torres Strait region*, GBRMPA Workshop Series 16, AGPS, Canberra.

Lea JP, 1993. Regional development planning in the Torres Strait: new directions in Melanesian Australia, Paper prepared for the Western Regional Science Association Meeting, Maui, Hawaii, February 1993.

Lea JP, Stanley OG & Phibbs PJ, 1990. *Torres Strait regional development plan 1990–95*, Department of Urban and Regional Planning, University of Sydney, Sydney.

Legislation Review Committee, 1991. *Towards self-government*, Discussion paper of the inquiry into legislation relating to the management of Aboriginal and Torres Strait Islander Communities in Queensland, Cairns.

Mulrennan ME, 1992. *Coastal management: challenges and changes in the Torres Strait Islands*, Discussion Paper No.5, North Australia Research Unit, Darwin.

Mulrennan ME & Jull P, 1992. Indigenous people and sustainability: the politics of regional development in Australia's Torres Strait and Canada's Nunavut, conference paper for The Association for Canadian Studies in Australia and New Zealand (ACSANZ), Wellington, December 1992.

Murray R, 1992. The assertion of rights by Torres Strait Islanders under the Torres Strait Treaty, unpublished dissertation, Faculty of Law, University of New South Wales, Sydney.

O'Rourke DJ, 1988. *Report of the Interdepartmental Committee on the Torres Strait Islands*, Department of Aboriginal Affairs, Canberra.

Scott JM, 1990. Torres Strait independence: issues in island development, *in* Babbage, Annex R, 383–409.

Smyth D, 1992. *A voice in all places: Aboriginal and Torres Strait Islander interests in Australia's coastal zone*, Draft consultancy report for the Coastal Zone Inquiry, Resource Assessment Commission, November 1992.

Torres Strait Protected Zone Joint Authority TSPZJA, 1990. *Annual Report*, Dept of Primary Industries and Energy, AGPS, Canberra.

Van Beurden J, 1992. The role of Aboriginal and Torres Strait Islander peoples in coastal management, *AIC New Directions in Coastal Management Conference*, Sydney, 1–22.

Mabo case

Allen D, 1993. Salt-water dreaming, *in* Jull, Mulrennan & Sullivan (eds), *Surviving Columbus: indigenous environments today and tomorrow*, Proceedings of conference organised by the NARU, ICC, NLC and CLC, October 1992, forthcoming.

Brennan F, 1992. *Mabo and its implications for Aborigines and Torres Strait Islanders*, Uniya, Sydney.

Davis SL & Prescott JRV, 1992. *Aboriginal frontiers and boundaries in Australia*, Melbourne University Press, Melbourne.

Jull P, 1992. Indigenous rights: after the *Mabo* case, *Arena*, No.99/100, 61–4.

Sharp N, 1990. Contrasting cultural perspectives in the Murray Island land case, *Law in Context* **8**(1), 1–31.

Sharp N, 1991. A landmark: the Murray Island case, *Arena*, No.94, 78–93.

Sutherland J, 1992. Rising sea claims on the Queensland east coast, *Aboriginal Law Bulletin* **2**(56), 17–19.

Wright J, 1991. A shaft of light, *Island*, No.47, 44–7.